高校专门用途英语（ESP）系列教材

ENGLISH for SOFTWARE PROJECTS
Reading & Writing

软件项目英语 读写

张宏岩 编著

清华大学出版社
北京

内 容 简 介

本书通过案例情景教授软件工程师如何使用英语完成独立软件开发的全流程任务，采用职场沟通场景和软件开发场景双线索的方式，从新入职开始逐步经历项目可行性研究、需求分析、设计、详细设计、编码、测试到最终项目总结，覆盖各阶段所需职场沟通和技术对话的英语听说技能。适合软件类专业的大学本科生、硕士生和需要使用英语从事软件开发的从业者。

练习答案、教学资源可在ftp://ftp.tup.tsinghua.edu.cn/上下载。

版权所有，侵权必究。举报：010-62782989，beiqinquan@tup.tsinghua.edu.cn。

图书在版编目（CIP）数据

软件项目英语：读写 / 张宏岩编著. —北京：清华大学出版社，2021.8（2023.7重印）
高校专门用途英语（ESP）系列教材
ISBN 978-7-302-50109-1

Ⅰ.①软… Ⅱ.①张… Ⅲ.①软件工程—英语—阅读教学—高等学校—教材
②软件工程—英语—写作—高等学校—教材　Ⅳ.①TP311.5

中国版本图书馆CIP数据核字（2018）第106250号

责任编辑：钱屹芝
封面设计：子　一
责任校对：王凤芝
责任印制：宋　林

出版发行：清华大学出版社
网　　址：http://www.tup.com.cn, http://www.wqbook.com
地　　址：北京清华大学学研大厦A座　　邮　编：100084
社 总 机：010-83470000　　邮　购：010-62786544
投稿与读者服务：010-62776969，c-service@tup.tsinghua.edu.cn
质量反馈：010-62772015，zhiliang@tup.tsinghua.edu.cn

印 装 者：天津鑫丰华印务有限公司
经　　销：全国新华书店
开　　本：170mm×230mm　　印　张：19.25　　字　数：295千字
版　　次：2021年8月第1版　　　　　　　　印　次：2023年7月第2次印刷
定　　价：78.00元

产品编号：080082-01

序

　　说起软件，不能不提到印度。我在2006年撰写的《大学英语教学：回顾、反思和研究》一书中，曾这样评论印度人的软件神话："自20世纪80年代以来，为了适应世界的高科技需求，印度迅速地、甚至奇迹般地发展了自己的软件产业，在短短的十几年里，软件生产量已经占世界软件总量的16.7%，成为世界第二软件大国，软件科技人员总数仅次于美国而居世界第二，其信息技术知识产权出口以年均35%的高速度增长。"美国微软巨头比尔·盖茨曾惊呼"印度将会在21世纪成为超级软件大国"。

　　印度之所以能够做到这一点，我当时分析，很重要的一个原因就是他们软件从业人员英语纯熟，能用英语熟练摄取行业里最前沿的信息和科研成果。中国软件从业人员或工程师人数不比印度同行少，技术也不差。但是比起他们用英语汲取信息和国际交流的能力，我们的弱点就暴露了出来。2005年麦肯锡调查报告指出，"中国目前工科毕业生中只有不到10%的毕业生具备在国际环境下工作所需要的语言技能"，英语水平高的软件人才更是凤毛麟角，可以说，工程师的英语能力已成为制约我国软件产业国际竞争力的一个重要因素。

　　英语能力可以分为通用英语能力和专门用途英语（ESP，English for Specific Purposes）能力，两者是有差异的。传统的观点是只要基础扎实就能胜任专业方面的工作和研究，这是一个误区。任何一个学科的理论知识构建和传播都是通过特定的语言方式来实现的。如自然学科往往通过图表公式来描述实验结果；人文学科则通过引经据典来阐述一个观点。因此，即使同一语类如期刊论文、项目报告等，其语篇结构、修辞手段、句法特征也是因学科而异，其话语范式和语言表达是因行业而异的。因此，任何学生或工程师，不管他们英语语言基础如何，在他们专业圈子或者行业共

I

同体里都是一个新成员。只有通过结合他们专业或行业的独特的话语范式、典型工作场景进行专门的语言训练，才有可能融入他们各自的学科或行业共同体内，用英语开展有效的工作和研究。

最近几年，我国出版的专门用途英语教材不少，但真正符合 ESP 理念的教材不多，好多仅仅是专业英语教材的翻版，或是全英语专业教材的通俗本。我们必须清楚，专业英语主要是介绍某个学科或行业术语、概念、理论等，而专门用途英语则主要分析特定学科或行业里的知识和理论是用什么方法构建和交流的，主要是训练学生和科技人员在特定行业里的英语交流能力，两者之间有很大差异。张宏岩博士主编的这本教材让我眼睛一亮，这是最接近我理想中的专门用途英语教材。

这本教材围绕软件项目的启动、策划和实施等专业内容展开，并且具有下面 ESP 的特征：（1）内容直接来自行业的实践，符合软件行业典型的工作场景和流程；（2）教材所培养的能力是当前软件行业实践最紧缺的技能，比如软件行业所需的文档阅读能力；（3）教材所选材料能较好反映英语语言在该特定领域的使用规范（如真实可借鉴的软件文档）和使用规律，所涉及的情境具有可扩展性，能够让读者举一反三。

能编写出这样一本理想的专门用途英语教材和张博士多学科背景、在北京大学软件与微电子学院的教学经历、在工业和信息化部教育与考试中心 IT 职业英语水平考试项目的有关工作经历有关。尤其是，张博士作为我国"专门用途英语专业委员会"的常务理事，做了大量开拓性的实践工作，积累了非常宝贵的经验，这一切都为此书的编写和出版奠定了良好的基础。真心希望该教材能为我国培养具有国际竞争力和国际话语权的软件精英做出贡献，为我国专门用途英语教材的建设提供可借鉴的途径。

蔡基刚
复旦大学教授、博导
中国英汉语比较研究会
专门用途英语专业委员会会长
2020 年 8 月

前言

在我从事英语教育的20多年中,最难忘记的是一个学生跟我描述的他和英语的悲情故事。这位同学因为英语不够好在IBM软件工程师的面试中败下阵来,于是加入了联想。然而2004年联想并购了IBM的PC和笔记本电脑部,他刚有起色的IT人生又遭到尾随而至的语言挑战,他说最尴尬的人生一幕是:在一次有关中国区业务的讨论中,他听到一位美国同事对中方人员毫无根据地指责,他忍无可忍,拍案而起,大喊了一声"NO!"。当激动的中国同事们带着期望看向他时,他居然卡壳了,"我大脑运行的速度远远超出了舌头能够反应的速度,那一刻,学了多年的英语关键时刻竟没能再冒出一个单词来,无语凝噎!"他身上所反映出的是九成半传统英语课程体系下成长起来的软件人员的普遍情况:所学的英语和行业应用无法接轨,更确切地说,在大学和其后的英语学习中,从未接受过如何将英语应用在软件项目工作中的训练,这是一个系统性的缺失。

2005年,笔者作为工信部教育与考试中心全国IT职业英语水平考试项目组组长,组织了20人的专家团队,陆续对全国40家IT类外企、合资企业、软件外包企业的200余名高级软件工程师、软件项目经理、软件开发与测试人员的外语需求做了调研,调研发现:软件项目过程中,一个合格的工程师或项目经理应具备:(1)一定的商务英语口语沟通能力;(2)软件开发过程中使用技术英语沟通的口语能力;(3)高效、准确阅读技术文档的阅读能力;(4)编写软件开发各阶段技术文档或报告的写作能力。一言以蔽之,所缺的就是软件项目英语应用能力。

2006年至2008年,我先后在北大、北航、北理工、哈工程这四所高校的软件学院教授软件英语类课程,同一时期获得了中关村科技园区管委会软环境建设专项基金资助,研发适合中国软件工程师的英语课程体系,

本书的早期版本是一个研究成果，并在清华科技园试点培训软件工程师，为 SUN、MOTOROLA、联想、紫光、软通动力等企业的软件项目经理和工程师提供了专题培训。课程体系进而吸纳了不同领域的专家组提出的中肯意见，融合了计算机教育、英语教育、国际 IT 企业内训三个领域的教育理念。2013 年，我开始在北京大学软件与微电子学院全职任教，经过在教学实践中对本书核心内容的进一步完善修订，这本凝聚了"帮助中国软件人才突破英语瓶颈"梦想的《软件项目英语：读写》如今终于有机会正式公开出版发行。

鉴于书中内容从取材、编撰到应用都已经在高校和企业当中得到过成功的实践，本书既适合作为高等院校软件学院本科和研究生层次的专业英语教材，也适合作为软件从业人员的内训或自学教材。

由衷感谢以下人员在本书初稿编辑和审校过程中所做出的贡献：Janice Willson（美）、Corrin Nielson（美）、Charles Odenhal（美）、Robert Makelin（美）、Christine Hansell（美）、Brian Connors（美）、段磊、冯宇、李子亮、韩清月、李健利、王秋桐、宋京晶、张露薇。感谢 Janice Willson 和 Robert Makelin 精心录制的配套音频。感谢协助本书修订工作的尹玉珺。感谢资深国际软件开发专家 Gerald Cheong、张玉超、Eric Dao、李扬等为确定国际软件企业英语应用需求及确立应用规范提供的直接帮助。最后还要感谢众多参与了需求调查和章节测试的中关村软件园区企业的软件工程师们，他们的反馈意见提高了教材的针对性、互动性，最终使教材更好地贯彻了以用户体验为主导的编撰思路。

祝愿所有的本书学习者都能够在这个虚拟的全流程软件项目开发过程中汲取所需的各样养分，成为全球化时代精通英语的中国软件精英。对本书有任何意见和建议，欢迎您联系编著者 drzhy@pku.edu.cn。

张宏岩

2020 年 8 月 28 日

Contents

Unit 1 Getting Started 项目启动 1

I Feasibility reports (Reading)

An overview; Reading a feasibility report; Post-reading exercises

II Communicating by e-mail (Writing)

Addressing an e-mail correctly; Explaining a feasibility report; Guide to technical writing; Technical writing exercises

Unit 2 Formulating Plans 项目策划 31

I Project plans (Reading)

An overview; Reading a software development plan; Post-reading exercises

II Discussing by e-mail (Writing)

Writing meaningful subject lines; Discussing by e-mail; Guide to technical writing; Technical writing exercises

Unit 3 Specifying Customer Needs 明确需求 65

I Software requirements specifications (Reading)

An overview; A sample of software requirements specification; Post-reading exercises

II E-mail & technical writing (Writing)

Salutation, greeting & identification; E-mails about software functions; Guide to technical writing; Technical writing exercises

Unit 4 Describing Designs 设计简述 89

I Software design documents (Reading)

An overview; A sample of software design document; Post-reading exercises

II E-mail & technical writing (Writing)

The length, content and format of the message; E-mails about design details; Guide to technical writing; Technical writing exercises

Unit 5 Detailing Processes 操作细则 123

I UML & design specifications (Reading)

An overview; An architectural design specification; A sample of detailed design specification; Post-reading exercises

II E-mail & technical writing (Writing)

Manner & tone; E-mails describing UML diagrams; Guide to technical writing; Technical writing exercises

Unit 6 Documenting Your Work 文档制作 157

I Source code documentation conventions (Reading)

An overview; A naming and code documentation guide; Post-reading exercises

II E-mail & technical writing (Writing)

Replying or forwarding e-mails properly; E-mails about software documentation; Guide to technical writing; Technical writing exercises

Unit 7 Implementing a Project 项目实施 181

I Project work plans (Reading)

An overview; A sample of project work plan; Post-reading exercises

II E-mail & technical writing (Writing)

Using abbreviations and smileys; Exercises on e-mail writing; Guide to technical writing; Technical writing exercises

Unit 8 Negotiating Assignments 协商任务 213

I Unit test plans (Reading)

An overview; A sample of unit test plan; Post-reading exercises

II E-mail & technical writing (Writing)

Signing your e-mails; Exercises on e-mail writing; Guide to technical writing; Technical writing exercises

Unit 9 Testing Software 软件测试 241

I Software test plans (Reading)

An overview; A sample of software test plan; Post-reading exercises

II E-mail & technical writing (Writing)

Dealing with attachments appropriately; Exercises about e-mails and software defects; Guide to technical writing; Technical writing exercises

Unit 10 Closing Off 项目总结 271

I Post-mortem reports (Reading)

An overview; A sample of post-mortem report; Post-reading exercises

II Summaries (Writing)

Summary of guidelines to professional e-mails; E-mail about course summary; Guide to technical writing; Technical writing exercises

Unit 1

Getting Started

项目启动

I   Feasibility reports
II Communicating by e-mail

1 Feasibility reports
Reading

A An overview

Feasibility Study

A feasibility study is an analysis of a problem to determine if it can be solved effectively given the budgetary, operational, technical and schedule constraints in place. The results of the feasibility study determine which, if any, of a number of feasible solutions will be developed in the design phase. The aim of the feasibility study is to identify the best solution under the circumstances by identifying the effects of this solution on the organization.

Within the system development cycle, the feasibility study is undertaken after the problem has been defined and analyzed, but before undertaking detailed design of a solution. Defining the problem has quantified the needs, the objectives and the boundaries of the problem. This, to a significant extent, identifies the constraints.

The systems analyst usually undertakes the feasibility study. Sometimes, CTO or project manager may play the role of system analyst.

Based on analysis of the problem, presented in the Requirements Definition Report or User Requirements Document, the report writer uses his or her understanding of software design and development to describe and evaluate a feasible solution to the problem. Commonly a number of feasible solutions are described and evaluated. These are presented to management as alternatives or options in a Feasibility Report to allow management to select the best solution. There are three types of feasibility report:

The first type studies a situation (for example, a problem or opportunity) and a plan for doing something about it and then determines whether that plan is "feasible"—which means determining whether it is technologically possible and practical (in terms of current technology, economics, social needs, and so on). This type of feasibility report answers the question "Should we implement Plan X?" by stating "yes" "no", but more often "maybe". Not only does it give a recommendation, it also provides the data and the reasoning behind that recommendation.

The second type starts from a stated need, a selection of choices, or both and then recommends one, some, or none. For example, a company might be looking into grammar-checking software and want a recommendation on which product is the best. As the report writer on this project, you could study the market for this type of application and recommend one particular product, a couple of products (differing perhaps in their strengths and their weaknesses), or none (maybe none of them is any good). This type answers the question "Which option should we choose?" (or in some cases "Which are the best options?") by recommending Product B, or maybe both Products B and C, or none of the products.

The third type provides an opinion or judgment rather than a yes-no-maybe answer or a recommendation. It provides a studied opinion on the value or worth of something. This type of feasibility report compares a thing to a set of requirements (or criteria) and determines how well it meets those requirements. (And of course there may be a recommendation—continue the project, abandon it, change it, or other possibilities.)

B Reading a feasibility report

Read the following feasibility report. For the first time, please only scan the whole document. Keep these questions in mind and try to answer them after scaning. Time limit: 10 minutes.

- What is the major task of this proposed report?
- How many optional plans are mentioned here?
- Which option is finally recommended?

Feasibility Report, Planet Tracking Software GCC Corporation

Table of Contents

1. Document Overview

   1.1 Introduction

   1.2 Scope

2. Study of the Current and Desired Environment

   2.1 Present Environment

   2.2 Problems with Present Environment

   2.3 Desired Environment

3. Planet Tracking Software Requirements

4. Constraints

   4.1 Technical Constraints

   4.2 Financial Constraints

   4.3 Schedule Constraints

5. Possible Solutions

   5.1 Solution 1

     5.1.1 Operation Feasibility

     5.1.2 Technical Feasibility

     5.1.3 Schedule Feasibility

     5.1.4 Financial Feasibility

     5.1.5 Legal Feasibility

   5.2 Solution 2

     5.2.1 Operation Feasibility

Glossary

feasibility *n.* 可行性

tracking *n.* 跟踪；探测

scope *n.* 范围

constraint *n.* 限制, 制约

legal *adj.* 法律上的

5.2.2 Technical Feasibility
5.2.3 Schedule Feasibility
5.2.4 Financial Feasibility
5.2.5 Legal Feasibility
5.3 Solution 3
 5.3.1 Operation Feasibility
 5.3.2 Technical Feasibility
 5.3.3 Schedule Feasibility
 5.3.4 Financial Feasibility
  5.3.4.1 Costs
   5.3.4.1.1 Tangible Costs
    5.3.4.1.1.1 Hardware Costs
    5.3.4.1.1.2 Software Costs
    5.3.4.1.1.3 System Design and Analysis Costs
    5.3.4.1.1.4 Administrative Costs
    5.3.4.1.1.5 Time Costs
    5.3.4.1.1.6 Training Costs
   5.3.4.1.2 Intangible Costs
  5.3.4.2 Benefits
  5.3.4.3 Cost-Benefit Analysis
 5.3.5 Legal Feasibility
6. Conclusion
Appendices

> **Glossary**
> tangible *adj.* 有形的
> administrative *adj.* 管理的
> intangible *adj.* 无形的

1. Document Overview

1.1 Introduction

This document is a report on the feasibility study conducted on the proposed solutions for the Planet Tracking Software. This software will interface with the Planet Tracking Unit—a device that is being built by the Atmospheric Physics Lab of Cosmos Engineering. The feasibility study was conducted between June 15th, 20xx and June 21st, 20xx by Group 101, GCC Corporation.

This feasibility study starts with a study of the current environment, the problems within the current environment and a summary of the proposed environment. The functionality expected from the Planet Tracking Software is summarized in the next section. The detailed high-level requirements document for the Planet Tracking Software can be obtained at *www.cosmoseng.com*. The general constraints on the development process are summarized in the section that follows.

Section 5 examines all the proposed solutions. For each of the proposed solutions, five kinds of feasibility study are conducted—operational, technical, schedule, financial and legal.

An operational feasibility study examines how the software will change the roles of the stakeholders and the users and whether the new workflow and organizational structure will be accepted by the users and stakeholders.

Glossary

propose *v.* 提议
interface *v.* 连接
atmospheric *adj.* 大气的
conduct *v.* 进行
functionality *n.* 功能性
stakeholder *n.* 股东
workflow *n.* 工作流程

A technical feasibility study checks to see if the proposed solution is feasible given the skills of our group and the environment the software is expected to be deployed in the Planet Tracking Software.

A schedule feasibility study checks if the proposed solution can be developed in a manner that will ensure that all deadlines set by Cosmos Engineering and other clients are met.

A financial feasibility study examines the costs and benefits of developing the software in the manner of the proposed solution.

A legal feasibility study determines whether any infringement, violation, or liability that could result from development of the system.

This study then concludes with the proposed solution that is determined to be the most feasible alternative given the constraints.

1.2 Scope

This document covers the feasibility of the Planet Tracking Software. It does not cover the feasibility of the device the software is being built for—the Planet Tracking Unit. Throughout the document, it is assumed that the Planet Tracking Unit is feasible and the specifications of the unit provided to us by our client, Cosmos Engineering, is complete and accurate.

(2. Omitted)

Glossary

deploy *v.* 配置,使用
manner *n.* 方法,方式
deadline *n.* 最终期限
alternative *n.* 选择,选项
specification *n.* 技术说明
accurate *adj.* 准确的,精确的
omit *v.* 省掉,略去

3. Planet Tracking Software Requirements

The following is a summary of the features that will be required for the Planet Tracking Software to ensure the environment proposed in the previous section is achieved.

- The software should be able to control the motors such that the mirrors and the unit can be aligned in all desired configurations. The alignment achieved should allow light from the desired source to enter the aperture of the spectrometer.

- The software should provide a GUI interface that allows the user to select the type of measurement that is desired from the unit. There are 5 modes the software should be able to operate in to accommodate the different types of measurements that are taken at Cosmos Engineering.

(the 5 modes are omitted)

4. Constraints
4.1 Technical Constraints

The technical constraints are a result of the environment at the Atmospheric Physics Lab, the skills of the members of the GCC group and the deadlines that have to be met for this project.

The first technical constraint that is considered for coming up with the possible solutions is that

Glossary

align *v.* 排列; 调准
aperture *n.* 孔, 洞; 光圈
spectrometer *n.* 光谱仪
GUI 图形用户界面
accommodate *v.* 适应, 调节

the system has to work on the Windows operating system. The other technical constraint is a result of the skills of our group. Most of the members of our group are familiar with Java and C++. Therefore, these are the languages that are considered as candidates for the development of the software.

(4.2 and 4.3 are omitted)

5. Possible Solutions

Our group examines 3 possible ways of building the Planet Tracking Software given the technical, financial and schedule constraints outlined in the previous section. For each of the proposed solutions, a detailed feasibility study is conducted to determine which of them would be the most feasible alternative.

5.1 Solution 1

The first solution examined is to build the Planet Tracking Software entirely in Java. This solution means that we will have to write the code for generating the co-ordinates of a celestial object given an instance of time.

The advantage of this solution is that all 5 members of the group are well-versed in the programming language chosen. The software built using this solution would work on all platforms supported by the Sun JVM.

The biggest disadvantage of using this solution is that the celestial objects supported will be

---

Glossary

candidate *n.* 候选人；候补物

entirely *adv.* 完全地

generate *v.* 生成

well-versed *adj.* 非常精通的

JVM = Java Virtual Machine 虚拟机

solar *adj.* 太阳的

restricted to just the required ones—the 9 planets of the solar system, the Sun and the Earth's moon. The following is the feasibility study for this alternative.

5.1.1 Operation Feasibility

The user of the Planet Tracking Software will be the person currently employed by the Atmospheric Physics Lab, Cosmos Engineering for taking the measurements. The advent of the Planet Tracking Unit together with the implementation of the Planet Tracking Software will significantly change the role of the user. As outlined in the Section 2.1 and Section 2.2, a large amount of the user's time is spent in aligning the mirror of the spectrometer. Since the alignment of the mirrors and the unit will be automatically done by the Planet Tracking Software, this activity of the user will be eliminated. Once the software is implemented, the user will be expected to spend most of his/her time in the analysis of the readings given by the spectrometer.

5.1.2 Technical Feasibility

All five of the members of our group are familiar with Java. Hence this solution does not translate into a learning curve for any of the members. There are two modules of the Planet Tracking Software that have been identified as critical sections of the code that would be required for the software. The first critical section is where the code will communicate with the motors using the Prairie Digital Model 40 over the serial port. Java code that performed a

Glossary

advent *n.* 到来, 出现
implementation *n.* 执行
eliminate *v.* 排除, 消除
learning curve 学习曲线
serial port 串行端口

portion of the functionality required by this module is written by one of the members to ensure the feasibility of this section.

The second critical section is where the code determines the co-ordinates of a celestial object using the time provided. We have not written Java code that exhibits any of the functionality required by this module yet. But our group does not anticipate this to be a problem as there are analogous pieces of code written in C++ and converting from C++ to Java is not expected to be a problem.

(5.1.3, 5.1.4 and 5.1.5 are omitted)

5.2 Solution 2

The second solution examined is to build the Planet Tracking Software entirely in Visual C++. This solution, like Solution 1, means that we will have to write the code for generating the coordinates of a celestial object given an instance of time.

The advantage of this solution is that the software would be highly stable due to the fact that it will use only the platform SDK. The software built will have the Windows look and feel, which will make the user comfortable with the interface, as the users at the Atmospheric Physics Lab primarily use the Microsoft Windows family of operating systems.

The biggest disadvantage of using this solution, e.g. Solution 1, is that the celestial objects supported will be restricted to just the required ones—the nine planets of the solar system, the Sun and the

> **Glossary**
>
> exhibit *v.* 展出，显示
> anticipate *v.* 预期，预料
> analogous *adj.* 类似的
> SDK = Software Development Kit 软件开发工具包
> dependency *n.* 依赖

Earth's moon. The platform dependency is not a problem because all the client computers use the Windows family of operating systems and even the spectrometer software that is used to extract the required readings from the spectrometer is written for Microsoft Windows. The following is the feasibility study for this alternative.

5.2.1 Operation Feasibility

The operational feasibility study of this solution does not differ from the one outlined in Section 5.1.1 for the Solution 1.

5.2.2 Technical Feasibility

Only one member of our group is familiar with Visual C++. All the five members of our group are proficient in C++. Writing code in Visual C++ is the same as writing code in C++ except in one module. This module is responsible for drawing the GUI, and the code for this module is Visual C++ specifically. Since coding of this module would require the participation of only two members, this solution translates into a learning curve for only one member of our group.

There are two modules of the Planet Tracking Software that have been identified as critical sections of the code that would be required for the software. The first critical section is where the code will communicate with the motors using the Prairie Digital Model 40 over the serial port. Visual C++ code that performs all of the functionality required by this module has already been written

Glossary

extract *v.* 得到，获得

by our group. The second critical section is where the code determines the co-ordinates of a celestial object using the time provided. Although we have not written the entire module, we have written parts of this module to ensure that this section of our software is feasible.

5.2.3 Schedule Feasibility

None of the members of the group have dealt with astronomical subjects before. Hence there is a learning curve that is constantly present through all the phases of development of the Planet Tracking Software. This solution requires one member of our group to familiarize himself with Visual C++ syntax. Although this is an additional learning curve, it is not a steep one given the coding requirements and the programming experience of the concerned member.

We conclude that, given the deadlines established for the project by Cosmos Engineering, and our knowledge and experience, we can meet all deadlines established throughout the development process of the Planet Tracking Software.

5.2.4 Financial Feasibility

The members of our group already have a licensed copy of the Microsoft Visual Studio that is required to develop the code in Visual C++. The only other requirement for this solution is that the software built with this alternative would run only on the Microsoft Windows family. All of the client computers are equipped with the required operating

Glossary

astronomical *adj.* 天文的
familiarize *v.* 使熟悉
steep *adj.* 陡峭的; 不合理

system. It follows that the costs and benefits this solution <u>entails</u> are no different from the ones outlined in Section 5.1.4 for Solution 1, as the change in choice of programming language has not added any costs to the software.

5.2.5 Legal Feasibility

Just as in Solution 1, there are no legal issues with this solution. All software that is used to develop this software is legally purchased. All <u>algorithms</u> that are used in the software, especially the algorithms used to generate the position of celestial objects, <u>are under public domain</u> and can be used as required.

5.3 Solution 3

The third solution examined was to build the Planet Tracking Software in Visual C++ but not in its <u>entirety</u>. Given an instance of time, Starry Night would be used to generate the co-ordinate information for a celestial object. The solution would involve writing a plug-in for Starry Night using the SDK provided by IMAGINOVA. Using this plug-in, the Planet Tracking Software would be able to obtain the co-ordinates for any celestial object that is supported by Starry Night.

The advantage of this solution is that we do not have to worry about the algorithms for generating the position of the celestial objects given a particular instance of time. This functionality of the software would be taken care of by the Starry Night software. Also, Starry Night supports over 28,000 <u>galaxies</u>

> **Glossary**
>
> entail *v.* 需要；伴随
> algorithm *n.* 算法
> be under public domain
>   无版权限制
> entirety *n.* 全部，全体
> galaxy *n.* 星系

and 1,000,000 stars in addition to our solar system (includes all the moons of all the planets in our solar system). The disadvantage of this solution is that the stability of the Planet Tracking Software would be highly dependent on the stability of Starry Night. Starry Night is an extremely large piece of software that contains a lot more functionality than just generating the positions of celestial objects. Our group has observed Starry Night to <u>crash</u> a lot of times while testing. Using this software to obtain our co-ordinates means that our code will have to <u>account for</u> all the <u>exceptions</u> that might occur due to the usage of Starry Night. The following is the feasibility study for this alternative.

5.3.1 Operation Feasibility

The operational feasibility study of this solution does not differ from the one outlined in Section 5.1.1 for the Solution 1.

5.3.2 Technical Feasibility

Only one member of our group is familiar with Visual C++. All five members of our group are proficient in C++. Writing code in Visual C++ is the same as taht in C++ with an exception in two modules. The code for the module is responsible for drawing the GUI and the code for the plug-in for Starry Night is Visual C++ specifically. Since coding both of these modules has to be fairly concurrent, the coding of these modules would require the participation of three members. It means that this solution translates into a learning curve

Glossary

crash v. 崩溃，失败
account for 解释，说明
exception n. 例外，意外

for two members of our group. There are two modules of the Planet Tracking Software that have been identified as critical sections of the code required for the software.

5.3.3 Schedule Feasibility

None of the members of the group have dealt with astronomical subjects before. Hence there is a learning curve that is constantly present through all the phases of development of the Planet Tracking Software. This solution requires two members of our group to familiarize themselves with Visual C++ syntax.

Although this is an additional learning curve, it is not a steep one given the coding requirements and the programming experience of the concerned members.

As is apparent from the Gantt chart, accommodating the time required to develop the plug-in leads to the conflict with the deadlines. The estimated time of completion extends to April 06, 20xx. That is 2 weeks beyond the deadline provided by Cosmos Engineering. It is still possible to accommodate this solution by having the group members start with the system analysis phase of the project early (during the winter break) and complete the low level requirements study. This will allow the development of the plug-in to start early and will also free up another group member to help speed up the development process of this module.

5.3.4 Financial Feasibility

5.3.4.1 Costs

5.3.4.1.1 Tangible Costs

5.3.4.1.1.1 Hardware Costs

For this solution, the only hardware required for the implementation is a computer. A computer with the required specifications is already present in the Atmospheric Physics Lab. The client intends to use the Planet Tracking

Software on that computer. As a result, there will be no hardware costs incurred for the implementation of the Planet Tracking Software.

5.3.4.1.1.2 Software Costs

For this solution, the Starry Night software package which costs approximately $60 would have to be bought on client expenses. Starry Night software works best on Windows. The computer that the client intends to use the Planet Tracking Software on already has Windows installed. Thus, the client will have to spend a total of approximately $60 for additional software.

5.3.4.1.1.3 System Design and Analysis Costs

Assuming each of the five people in the team spends 250 hours over the course of the seven months on this project and charges $25/hour/person, the system design cost will be $31,250. But since the Planet Tracking Software is being implemented by the GCC group as part of a 4th year project, the client will not incur any costs.

5.3.4.1.1.4 Administrative Costs

The client will incur no administrative costs as the Planet Tracking Software is going to be operated by a person already employed by them.

5.3.4.1.1.5 Time Costs

Over the time period involved in the development of the system, we estimate that the client will spend approximately 60 hours with us. Assuming a salary of $40/hour, our client will incur a cost of

Glossary

incur *v.* 招致，承受

approximately *adv.* 大约

approximately $3,000.

5.3.4.1.1.6 Training Costs

Since the user works at the Atmospheric Physics Lab, he will be familiar with all of the input and output that are associated with the Planet Tracking Software. The training will involve familiarizing the user with the software and showing the user how to customize the settings that the software uses. With the addition of the option of choosing more celestial objects and the Starry Night software, the training hours will increase in comparison to Solution 1 and Solution 2. We estimate that training for using this solution to build the software would take a maximum of 2 days (12 hours). Estimating a pay of $40/hour, this would translate to a cost of $480 for the client.

5.3.4.1.2 Intangible Costs

This solution, like all the proposed solutions so far, do not have many external dependencies. The system once installed will not require much maintenance or upgrades. The fact that the Planet Tracking Software does not generate any data that needs to be stored means that there are no costs that would accrue over the years.

5.3.4.2 Benefits

The benefits of the Planet Tracking Software using this alternative will be no different from the ones outlined in Section 5.1.4.2 for Solution 1.

5.3.4.3 Cost-Benefit Analysis

Glossary

external *adj.* 外部的

maintenance *n.* 维护

As seen in the feasibility study, the client will incur an expense of $3,540. 85% of this cost is a result of the amount of time the client is expected to spend with our group during the development of the Planet Tracking Software. This estimated expense is a one-time cost and no cumulative expenses have been identified to be a factor in this solution.

As there is no dollar figure that can be placed on the benefits, it is difficult to do a cost-benefit analysis. Our client ensures that this expense is worth the investment for the return in benefits that they would receive. Moreover, this is the most expensive solution seen so far that was proposed by the client initially.

5.3.5 Legal Feasibility

There are no legal issues identified with this solution. No copyrighted algorithms are expected to be used in this solution for the Planet Tracking Software.

The external software that will be used by the solution Starry Night will be bought. Writing plug-ins for Starry Night does not infringe on any copyrights either. The SDK for Starry Night is made available to the public, and IMAGINOVA, makers of Starry Night, encourages developers to write plug-ins for their flagship software.

6. Conclusion

By examining the results of the feasibility

> **Glossary**
> initially *adv.* 最初，开始
> flagship *n.* 旗舰，王牌

study, the GCC group has decided that the second alternative (Solution 2) is the most feasible one. The biggest factor that counts against Solution 3 is the reliance on Starry Night. None of the group members are <u>enthusiastic</u> about the idea of relying on a third-party software for information. Solution 3 also runs a risk of <u>overrunning</u> the deadlines established by Cosmos Engineering.

Solution 2 and Solution 1 are similar to each other in their feasibility studies. Both of them seem to indicate a process whereby all the deadlines will be met comfortably. Solution 1 is rejected because most of the members prefer programming in C++ to programming in Java. Since using Solution 2 to build the software only uses platform SDK, it also seems to be more stable and reliable solution.

Glossary

enthusiastic *adj.* 热情的
overrun *v.* 超过限度

C Post-reading exercises

Exercise 1

How many different types of feasibility study are conducted generally in a proposed solution? What are they?

Exercise 2

Fill in the table according to the report.

	Programming Language	Advantages	Disadvantages	Final Comments
Solution 1				
Solution 2				
Solution 3				

Exercise 3

Translate the following sentences into Chinese.

a) This document is a report on the feasibility study conducted on the proposed solutions for the Planet Tracking Software.

b) The software should be able to control the motors so that the mirrors and the unit can be aligned in all desired configurations.

c) The first technical constraint that is considered for coming up with the possible solutions is that the system has to work on the Windows operating system.

d) The advantage of this solution is that all five members of the group are well-versed in the programming language chosen.

e) As is apparent from the Gantt chart, accommodating the time required to develop the plug-in leads to conflict with the deadlines.

II Communicating by e-mail
Writing

A Addressing an e-mail correctly

E-mails are now considered an important means of communication. In the IT industry, where efficiency and brevity are appreciated, electronic communication has been increasingly popular because of its speed and broadcasting ability. Here is a series of guidelines to help you write effective professional e-mails.

Communicating Professionally and Effectively by E-mail

—Addressers & addressees

√ **Always provide a Personal Name if your mail system allows it.**

Some mailers allow you to provide a "Personal Name" which attaches to your e-mail address as a textual comment, usually displayed as, for example, *Li Lei [lilei@eptip.org]*. A Personal Name attached to your address identifies you better than your address alone.

√ **Put only the people you are directly addressing in the "To" field.**

The "To" field is for the people with whom you are directly communicating. If you enter every address in the "To" field when the e-mail is directed to only some of them, the recipient would have no clue who should take action.

√ **Enter the people you are indirectly addressing in the "Cc" field.**

"Cc" is short for Carbon Copy. If you add a recipient's name to the "Cc" field in a message, a copy of the message is sent to that recipient, and the recipient's name is visible to other recipients of the message. You may need to send a copy of an e-mail to someone just to inform him/her of information they might be interested in, but shouldn't have to take action on.

√   **Use "Bcc" when addressing a message that will go to a large group of people who don't necessarily know each other.**

"Bcc" is short for Blind Carbon Copy. Just as it is not appropriate to give out a person's telephone number without his or her knowledge, it may be inappropriate to give out someone's e-mail address. For instance, when you send an e-mail message to 30 people and use "To" or "Cc" to address the message, all 30 people see each other's e-mail addresses. By using "Bcc", each recipient sees only two— theirs and yours. However, using "Bcc" is somewhat unethical and therefore its use is discouraged in professional environments.

×   **Use inappropriate or inexplicit Personal Names.**

Don't use phrases such as "Guess who" or "Harry Potter". These phrases are annoying as Personal Names and hinder the recipient's quick identification of you and your message. Your mails may even get filtered out as junk mail by some mailing service providers.

×   **Use "Cc" to copy your message to everyone.**

This is particularly true at work. If only a few people really need to receive your message, direct it only to them. Try not to use the "Cc" field unless the recipient in the field knows why they are receiving a copy of the message. That is to say, do not include the person in the "Cc" field when you don't have a particular reason for wanting this person to see what you have written.

B Explaining a feasibility report

Read this e-mail written by Jacky to Mr. Roland before Jadey made a telephone call. This kind of e-mail is usually more formal than those we write to our friends. Learn this different style to make use of it later.

From: jackychen@generalcomputers.com
To: roland@cosmoseng.com
Cc: davidzhang@generalcomputers.com
Subject: Feasibility report
Date: Tue, Jun 21, 2005
Attachments: feasibility_report_PTS_final.pdf

Dear Mr. Roland,

  I am terribly sorry that I was unavailable when you called me this morning. I have received your message, and I apologize for the delay. We are doing everything we can to pick up the pace.

  At present, we are involved in the requirements analysis phase, which is a very important step in software engineering to make sure the later processes will be smooth and successful. We've done a lot of work on it, and tried our best to speed up the work.

  We have already finished the feasibility report, and I'm attaching that report to this e-mail. In the report, we analyzed three possible solutions, and gave relevant data and a recommendation for solution two as the optimum choice.

  This work took one more day than expected, but we'll have to build our project plan and SRS based on it. I believe you can understand the effort we have made in order to bring about a win-win situation.

  Have a nice day!

Sincerely yours,

Glossary

terribly *adv.* 很, 极, 十分
pace *n.* 速度, 进度
smooth *adj.* 平坦的, 顺利的
speed up 加速
attach *v.* 附上; 连接
win-win *adj.* 双赢的

Unit 1　Getting Started

Jacky Chen

Software Engineer

General Computers Corporation

Writing assignments

The following are two hypothetical business situations. Read them carefully. Then compose two e-mails to deal with the issues.

1. Imagine you are writing an e-mail to your partner to thank him for his support in a meeting. Your team is evaluating the feasibility of developing a professional music composer for a client, Mr. Wang. A meeting was held to discuss whether to use the plug-ins from a third party application—Adobe Audition. You argued against relying on third party applications, and one of your co-workers strongly agreed with you. Therefore:

Write him an informal e-mail to thank him;

Tell him why you insisted on not using 3rd party plug-ins;

Send the e-mail to your partner's e-mail address as well as your teacher's.

2. Write an e-mail to the client, Mr. Wang. When you write to a client, use a formal style, and always think about how he would react to your words. Do the following tasks:

Tell Mr. Wang your team's decision—you are not going to rely on any third party applications; rather you are going to develop a piece of music composing software yourselves;

Explain the reasons;

Persuade him to push back the deadline by two weeks;

Send your e-mail to your partner as well as your teacher, and ask them for comments;

Try to improve your style and word choice well enough to persuade your client.

Guide to technical writing

No matter what your current or future job is, writing will be essential to your work because you will have to communicate your technical knowledge to others. Technical documents are a most frequently used type of writing in the IT industry, so training in technical writing will help ensure a smooth career path. Here is a series of guidelines to help you write successful technical documents.

Guidelines to Successful Technical Writing

—Pre-writing

Technical writing is the activity that creates documentation for a technology. Its purpose is to effectively communicate a message. The writer is responsible for writing text that is helpful to its intended audience, accurate, readable, and accessible. In contrast with other types of writing, any technical document:

- is the product of a writer who fully understands the subject,
- focuses on the subject, not the writer,
- conveys one meaning,
- is tailored to the specific needs of an audience,
- is at a level of technicality that will be understood by the specified audience, and
- is efficient.

A key part in the definition of technical communication is the receiver of the information—the audience. Technical communication is the delivery of technical information to readers (or listeners or viewers) in a manner that is adapted to their needs, level of understanding, and background. The ability to translate technical information to non-specialists is a key skill to any technical communicator.

Technical communication is a fully professional field; however, the purpose for this guide is not necessarily to train you as a technical writer, but to introduce the concepts and writing skills you will need for most technically-oriented professional jobs. No matter what sort of professional work you do, either systems designing, or programming, or testing, you're likely to do lots of writing—and much of it technical in nature. The more you know about technical writing, the better you're likely to do your job. That will be good for the projects you work on, for the organizations you work in, and most importantly for your career.

Before you start producing your document, you must make sure that you do the following:

1. *Determine the objective of the document.* This is critical. If you fail to do this, you will almost certainly produce something that is unsatisfactory. Every document should have a single clear objective. Make the objective as specific as possible.

2. *Write down the objective.* The objective should be stated at the beginning of the document. Ideally, this should be in one sentence. For example, the objective of this document is to help students write well structured, easy-to-understand technical documents. If you cannot write down the objective in one sentence, then you are not yet ready to start writing.

3. *Always have a specific reader in mind.* Assume the reader is intelligent, but uninformed. It may be useful to state the reader profile at the beginning of your document. For example, the target readers for this document are primarily students and software engineers with a good working knowledge of English. The document is not suitable for children under 13, or people who have yet to write documents in English. It is ideal for people who have written technical or business documents and wish to improve their writing skills.

4. *Decide what information you need to include.* You should use the objective as your reference and list the areas you need to cover. Once you have collected the information, make a note of each main point and then sort them into logical groups. Ultimately you have to make sure that every sentence makes a contribution to the objective. If the material you write does not make a contribution, remove it—if it is good, you may be able to use it in a different report with a different objective.

5. *Have access to a good dictionary.* Before using a word that "sounds good", but whose meaning you are not sure of, check it in the dictionary. Do the same for any word you are not sure how to spell.

6. *Identify someone who can provide feedback.* Make sure you identify a friend, relative or colleague who can read at least one draft of your report before you submit it formally. Do not worry if the person does not understand the technical area—they can at least check the structure and style and it may even force you to write in the plain English style advocated here.

D Technical writing exercises

Exercise 1

Locate a brief example of a technical document (or section) in a library or elsewhere. Make a photocopy or hard copy, and bring it to class. Prepare to explain why your selection can be called technical writing.

Exercise 2

Revise the following sentences to make them more audience-friendly. The first one has been done for you.

a) I have scheduled the requirement analysis activities to begin on March 1st.

   Our team is expected to begin the requirement analysis activities on March 1st.

b) We're requesting all team members to complete the enclosed questionnaire so that we may develop a master schedule for the project.

c) We are planning an in-house training program for employees who want to improve their writing skills.

d) Our safety policy forbids us from renting the power equipment to anyone who cannot demonstrate proficiency in its use.

e) To prevent us from possibly going beyond the budget for our project, the shareholders now require verification of any large check presented for immediate payment.

Exercise 3

Imagine your team is investigating the feasibility of a project in which you are to develop a professional electronic music composer—Creative Studio. Two possible solutions were presented in a meeting: 1) your team will create the program and the effective filters; 2) your team is going to write the main part of the program, but will use effective filter plug-ins from other applications such as Adobe Audition and Fruity Loops Studio.

Refer to the conclusion section of the document in Section 1.3, and write a similar conclusion of feasibility study according to the above information. Keep it within 300 words. Be certain to include a recommendation for an optimum solution in your report.

Unit 2

Formulating Plans

项目策划

I   Project plans
II Discussing by e-mail

1 Project plans
Reading

A An overview

What Is a Project Plan?

The project plan is the controlling document to manage an Information Technology (IT) project.

The project plan describes the:

- interim and final deliverables the project will deliver,
- managerial and technical processes necessary to develop the project deliverables,
- resources required to deliver the project deliverables,
- additional plans required to support the project.

Why Create a Project Plan?

Documenting the decisions is essential. As you record, gaps appear and inconsistencies protrude. Creating the project plan usually requires hundreds of mini-decisions and these bring clarity to the project.

The project plan communicates the decisions to others. Often what we assume is common knowledge which is unknown to other members of the team. Fundamentally, the project manager's goal is to keep everyone progressing in the same direction and communication is essential to achieve this goal. The project plan makes communicating a lot easier.

The project plan is a critical set of documents that should meet a project manager's need of information that is data-based. The project plan is a wealth of information as well as a checklist. By reviewing the project plan, as often as is required, the project manager knows where the project is to identify what correction action or changes of emphasis or shifts in direction are needed.

It is the job of the project manager to develop a project plan and to accomplish it. Only the written project plan is precise and communicable. The project plan consists of documents on who, what, why, when, where, how and how much. The project plan encapsulates much of the project manager's work. If their comprehensive and critical nature is recognized in the beginning, the manager can approach them as friendly tools rather than annoying overhead.

B Reading a software development plan

Read the following software development plan. This is only a preliminary plan, but it sheds much light on what a software development plan is. For the first time, please only scan the whole document. Keep these questions in mind and try to answer them after scanning. Time limit: 10 minutes.

- What is this plan mainly talking about?
- Why is this Software Development Plan proposed?
- What is the main constraint of this project?

Course Registration System Software Development Plan

Version 0.9

Table of Contents

1. Introduction
   1.1 Purpose
   1.2 Scope
   1.3 Definitions, Acronyms and Abbreviations
   1.4 References

Glossary

registration *n.* 注册, 登记
acronym *n.* 首字母缩写词
abbreviation *n.* 缩写词
reference *n.* 参考资料

1.5 Overview

2. Project Overview

   2.1 Project Purpose, Scope and Objectives

   2.2 Assumptions and Constraints

   2.3 Project Deliverables

   2.4 Evolution of the Software Development Plan

3. Project Organization

   3.1 Organizational Structure

   3.2 External Interfaces

   3.3 Roles and Responsibilities

4. Management Process

   4.1 Project Estimates

   4.2 Project Plan

      4.2.1 Phase Plan

      4.2.2 Iteration Objectives

      4.2.3 Releases

      4.2.4 Project Schedule

      4.2.5 Project Resourcing

         4.2.5.1 Staffing Plan

         4.2.5.2 Resource Acquisition Plan

         4.2.5.3 Training Plan

      4.2.6 Budget

   4.3 Iteration Plans

   4.4 Project Monitoring and Control

      4.4.1 Requirements Management Plan

      4.4.2 Schedule Control Plan

Glossary

objective *n.* 目标, 目的
deliverable *n.* 交付品
evolution *n.* 发展
iteration *n.* 迭代
release *n.* 版本; 发布
resourcing *n.* 资源管理
staffing *n.* 人力配置
acquisition *n.* 获取
budget *n.* 预算
monitor *v.* 监控

4.4.3 Budget Control Plan

4.4.4 Quality Control Plan

4.4.5 Reporting Plan

4.4.6 Measurement Plan

4.5 Risk Management Plan

4.6 Close-out Plan

> **Glossary**
>
> close-out *n.* 结算
> glossary *n.* 术语表
> vision *n.* 前景，蓝图

1. Introduction

1.1 Purpose

The objective of this Software Development Plan is to define the development activities in terms of the phases and iterations required for implementing a computerized class registration system for Zhonghua University.

1.2 Scope

This Software Development Plan describes the overall plan to be used by General Computers Corporation for developing the Course Registration System for Zhonghua University. The details of the individual iterations will be described in the Iteration Plans.

1.3 Definitions, Acronyms and Abbreviations

See Glossary (4).

1.4 References

Applicable references are:

1) Course Registration System Vision, V1.0, General Computers Corporation.

2) Course Registration System Business Case, DRAFT, 2004, General Computers Corporation.

3) Course Registration System Stakeholder Requests Document, V1.0, 2004, General Computers Corporation.

1.5 Overview

This Software Development Plan contains the following information:

Project Overview—provides a description of the project's purpose, scope and objectives. It also defines the deliverables that the project is expected to deliver.

Project Organization—describes the organizational structure of the project team.

Management Process—explains the estimated cost and schedule, defines the major phases and milestones for the project, and describes how the project will be monitored.

2. Project Overview

2.1 Project Purpose, Scope, and Objectives

This Software Development Plan describes the overall plan to be used by General Computers Corporation for developing the Course Registration System for Zhonghua University. The details of the individual iterations will be described in the Iteration Plans.

2.2 Assumptions and Constraints

The system is intended to be the primary

Glossary

deliver v. 交付
milestone n. 里程碑
primary adj. 首要的

means of student registration for the 20xx Fall Term. Since course registration begins on Sept 1 20xx, the system must be fully available by this date.

2.3 Project Deliverables

The following artifacts will be produced during the project, and delivered to the maintenance organization.

- the Product, including:
  - executable released Deployment Units
  - Installation Artifacts
- End-Use Support Material (including Release Notes)
- Source Code (Implementation Elements)
- Test Suites
- Software Architecture Document
- Project-Specific Design and Implementation Guidelines
- Use Cases
- Supplementary Specification
- Glossary
- Vision

Other artifacts will be produced, as described in the project development case, but are not intended to be delivered to the maintenance organization.

2.4 Evolution of the Software Development Plan

The Software Development Plan will be

> **Glossary**
>
> means *n.* 方式、方法
> artifact *n.* 工件
> end-use *adj.* 终端使用的
> guideline *n.* 指南
> supplementary *adj.* 补充的

revised prior to the start of each Iteration phase.

3. Project Organization
3.1 Organizational Structure

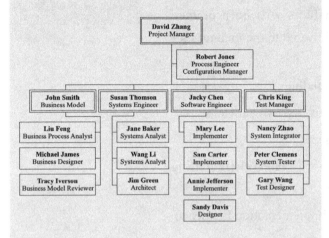

3.2 External Interfaces

The Project Manager will provide Status Assessment, as scheduled in this plan, to the IT Executive Stakeholder. The project team will also interact with other stakeholders to solicit inputs and review of relevant artifacts.

3.3 Roles and Responsibilities

The following table identifies the organizational units that will be responsible for each of the disciplines, workflow details, and supporting processes.

Glossary

status *n.* 状态

assessment *n.* 评估

executive *adj.* 执行的

solicit *v.* 索要

relevant *adj.* 有关的

Role	Responsibility
Project Manager	As described in the Rational Unified Process (6), responsible for managing the overall Project Management discipline. Leads the extended Project Management Team.
Process Engineer	Responsible for the project environment, and providing process related support for the teams in the project as defined in the Environment discipline in Rational Unified Process. Participates in an extended Project Management Team.
Configuration Manager/ Change Control Manager	Responsible for Configuration Control on the project, and for exercising the General Computers Change Request Process in the project. Participates in an extended Project Management Team.
Systems Engineering Team Leader	Leads the team primarily responsible for managing the Business Modeling and Requirements disciplines. Participates in an extended Project Management Team.
Software Engineering Team Leader	Primarily responsible for the Analysis & Design and the Implementation disciplines. Participates in an extended Project Management Team.
Test Team Leader	Leads the team responsible for managing the Test discipline. Participates in an extended Project Management Team.
Deployment Team Leader	Leads the team responsible for installation activities and infrastructure in the end-user environment. Participates in an extended Project Management Team.

> **Glossary**
>
> discipline *n.* 秩序
>
> infrastructure *n.* 基础设施

4. Management Process

4.1 Project Estimates

Project estimates are based on the Course Registration System Cost Model and Analysis

Report (7).

The Course Registration System is similar in complexity and architecture to the Online Library System, built for Zhonghua University in 1999. The course registration system database is roughly 25% more complex, and the number and complexity of use cases suggest that the system will be roughly 20% more complex overall. The time-frame and effort estimates from this report are the basis of the project budget and schedule.

4.2 Project Plan

4.2.1 Phase Plan

A Work Breakdown Structure is being prepared, and will be provided in the next version of this document.

The development of the Course Registration System will be conducted using a phased approach where multiple iterations occur within a phase. The phases and iterations in this plan do not overlap. A summary of the relative timeline is shown in the table below:

Phase	No. of Iterations	End
Inception Phase	1	Week 7
Elaboration Phase	1	Week 14
Construction Phase	1	Week 19
Transition Phase	4	Week 32

Glossary

complexity *n.* 复杂度
roughly *adv.* 粗略地
breakdown *n.* 分解
approach *n.* 步骤，途径
overlap *v.* 重叠
inception *n.* 开始，开端
elaboration *n.* 详细描述
construction *n.* 构架，建造
transition *n.* 交接

Unit 2 Formulating Plans

The following table describes each phase and the major milestone that marks the completion of the phase.

Phase	Description	Milestone
Inception Phase	The Inception Phase will develop the product require ments and establish the business case for the Course Registration System. The major use cases will be developed as well as the high level Software Development Plan. At the end of the Inception Phase, Zhonghua University will decide whether to fund and proceed with the project based upon the business case.	**Business Case Review Milestone** at the end of the phase marks the "Go/No Go" decision for the project.
Elaboration Phase	The Elaboration Phase will analyze the requirements and will develop the architectural prototype. At the completion of the Elaboration Phase, all use cases selected for Release 1.0 will have completed analysis & design. In addition, the high risk use cases for Release 2.0 will have been analyzed and designed. The architectural prototype will test the feasibility and performance of the architecture that is required for Release 1.0.	**Architectural Prototype Milestone** marks the end of the Elaboration Phase. This prototype of the major architectural components comprises the 1.0 Release.

Glossary

prototype *n.* 模型

(Continued)

Phase	Description	Milestone
Construction Phase	During the Construction Phase, remaining use cases will be analyzed and designed. The Beta version for Release 1.0 will be developed and distributed for evaluation. The implementation and test activities to support the R1.0 and R2.0 releases will be completed.	**Initial Operational Capability Milestone** (completion of the beta) marks the end of the Construction Phase.
Transition Phase	The Beta version for Release 1.0 will be distributed and evaluated. The Transition Phase will prepare the R1.0 and R2.0 releases for distribution. It provides the required support to ensure a smooth installation including user training.	**R2.0 Release Milestone** marks the end of the Transition Phase. At this point, all capabilities, as defined in the Vision Document (1), are installed and available for the users.

Glossary

capability *n.* 能力，权能

split *v.* 分开，拆分

illustrate *v.* 图解，说明

consist of 由……组成

subset *n.* 子集

Each phase is split into development iterations as described in Section 4.3.

Section 4.2.4 illustrates the high-level project schedule showing phases, iterations, and major milestones.

4.2.2 Iteration Objectives

Each phase consists of development iterations in which a subset of the system is developed. In general, these iterations:

- Reduce technical risk;
- Provide early versions of a working system;
- Allow maximum flexibility in planning features for each release;
- Enable scope changes to be handled effectively within an iteration cycle.

The following table describes the iterations along with associated milestones and addressed risks.

> **Glossary**
> flexibility *n.* 可适应性
> enable *v.* 使能够
> clarify *v.* 阐明
> realistic *adj.* 现实的
> mitigate *v.* 减低

Phase	Iteration	Description	Associated Milestones	Risks Addressed
Inception Phase	Preliminary Iteration	Defines business model, product requirements, Software Development Plan, and business case.	Business Case Review	Clarifies user requirements up front. Develops realistic Software Development Plans and scope. Determines feasibility of project from a business point of view.
Elaboration Phase	E1 Iteration—Develop Architectural Prototype	Completes analysis & design for all high risk requirements. Develops the architectural prototype.	Architectural Prototype	Architectural issues clarified. Technical risks mitigated. Early prototype for user review.

(Continued)

Phase	Iteration	Description	Associated Milestones	Risks Addressed
Construction Phase	C1 Iteration—Develop R1 Beta	Implement and test key R1 requirements to provide the R1 Beta Version. Assess if the release is ready to go for beta testing.	Initial Operational Capability (R1 Beta Code Complete)	All key features from a user and architectural perspective implemented in the Beta.
Transition Phase	T1 Iteration—Develop/ Deploy R1 Release	Deploy the R1 Beta. Fix defects from Beta, and incorporate feedback from Beta. Implement and test remaining R1 requirements.	R1 Beta Test Complete R1 Code Complete R1 Product Release	User feedback prior to release of R1. Product quality should be high. Defects minimized. Cost of quality reduced. Two-stage release.
Transition Phase		Package, distribute, and install R1 Release. Remaining low-risk R2 use cases fully detailed.		Minimizes defects. Two-stage release provides easier transition for users. R1 fully reviewed by user community.

Glossary

assess *v.* 评估

perspective *n.* 角度, 视角

Unit 2 Formulating Plans

(Continued)

Phase	Iteration	Description	Associated Milestones	Risks Addressed
Transition Phase	T2 Iteration—Develop R2 Internal 1	Design, implement, and test R2 Internal 1 requirements. Incorporate enhancements and defects from R1. Deploy the R2 Internal 1.	R2 Internal 1 Test Complete	If needed, R2 Internal 1 could be released to address R1 defects, to help address customer satisfaction.
	T3 Iteration—Develop R2 Internal 2	Design, implement, and test R2 Internal 2 requirements. Incorporate enhancements and defects from R2 Internal 2. Deploy the R2 Internal 2.	R2 Internal 2 Test Complete	R2 Internal 1 informally reviewed by user community. If needed, R2 Internal 1 could be released to address R1 defects, to help address customer satisfaction.
	T4 Iteration—Develop/Deploy R2 Release	Package, distribute, and install R2 Release.	R2 Code Complete / R2 Product Release	R2 Internal 2 informally reviewed by user community. Two-stage release provides easier transition for users.

Glossary

incorporate *v.* 整合

enhancement *n.* 改进, 增进

defect *n.* 缺陷

satisfaction *n.* 满意

target *v.* 定目标

4.2.3 Releases

This Software Development Plan addresses the first two releases of the Course Registration System. Key features as defined in the Vision Document (1) are targeted for the first 2 releases. All features critical to student registration are planned for the first release (R1.0).

The planned content of the releases is expected to change as the project progresses. This may be due to a number of business and technical factors. To accommodate the changes, Rational Requisite Pro will be used to manage the product requirements and to keep track of release content. In particular, benefit, effort, and risk attributes are used to determine priority of product requirements and thus the target release.

It is anticipated that the Course Registration System will be released for general use at Zhonghua University through 2 to 4 main releases.

Release 1 must contain at a minimum the basic functionality as listed below:

- Logon
- Register for courses
- Interface to Course Catalog Database
- Maintain Student Information
- Maintain Professor Information

Release 2 should include:

- Submit Student Grades
- View Grades
- Select courses to teach

The functionality for Release 3 has not yet been determined. It is anticipated that this release will contain enhancements to the existing functionality.

Glossary

priority *n.* 优先权
submit *v.* 提交

Future replacement of the legacy Billing System and Course Database System is targeted for Release 4 in Year 2007.

In addition, a Beta Release will precede the R1.0 Product Release, and will contain all the key R1 functionality. The Beta release will be deployed as if it were the real system, with the exception that it will interact with an isolated copy of the existing legacy systems, in order to avoid any disruption of existing systems. A select group of students and faculty will be asked to formally evaluate the beta.

In addition, there will be internal releases, to maintain a regular heartbeat to help keep the project on track, and to allow for the possibility of additional releases after the initial release, if needed. Internal releases can be informally reviewed by students and faculty. The following provides a brief description of the objectives for each of these internal releases:

- R2 Internal 1—addressing R1 bugs and enhancements. Implement main flow of "Select Courses to Teach", and design the UI for remaining R2 functionality.
- R2 Internal 2—implementing main flows for remaining R2 functionality.

4.2.4 Project Schedule

The high level schedule showing project phases, iterations, and milestones is contained in the Course Registration High Level Project Schedule (5) as shown below.

> **Glossary**
>
> legacy *n.* 旧有物
> precede *v.* 先于……出现
> avoid *v.* 避免
> disruption *n.* 中断
> faculty *n.* 教职人员
> formally *adv.* 正式地

	Date
Inception Phase	
Preliminary Iteration (start)	12/1/04
Business Case Review Milestone (end Inception Phase)	1/19/05
Elaboration Phase	
Iteration E1—Develop Architectural Prototype	
Architectural Prototype Milestone (end Elaboration Phase)	3/9/05
Construction Phase	
Iteration C1—Develop Release 1 Beta	
Initial Operational Capability Milestone (R1 Beta Code Complete)	4/9/05
Transition Phase	
Iteration T1—Develop/Deploy Release 1	
R1 Beta Test Complete	4/16/05
R1 Code Complete	4/30/05
R1 Product Release (end T1)	5/7/05
Iteration T2—Develop Release 2 Beta 1	
R2 Internal 1 Test Complete (end T2)	5/28/05
Iteration T3—Develop Release 2 Beta 2	
R2 Internal 2 Test Complete (end T3)	6/18/05
Iteration T4—Develop/Deploy Release 2	
R2 Code Complete	7/2/05
R2 Product Release (project close-out)	7/9/05

4.2.5 Project Resourcing
4.2.5.1 Staffing Plan

The IT employees identified in the Organizational Chart in Section 3.1 are allocated to the project. Additional resources will not be staffed until the business case is reviewed at the end of the Inception Phase and a Go/No Go decision is made on the project.

The test organization will rely on support from the software engineering organization, as shown by the dotted line in the organization chart.

4.2.5.2 Resource Acquisition Plan

The Zhonghua University IT department has insufficient Developers and Designers to meet the project needs. The Zhonghua University Recruiting Office is prepared to recruit a Senior Developer with several years C++ experience, and experienced System Integrator, and 2 Implementer/Testers (Junior Grade), with at least 1 year's C++ experience.

4.2.5.3 Training Plan

Training on the following skills will be conducted for the project team prior to the commencement of design activities:

- Object Oriented Analysis & Design
- Introduction to the Rational Unified Process
- Advanced C++ Features

Glossary

allocate *v.* 分派, 分配
insufficient *adj.* 不够的
recruit *v.* 招募, 补充
integrator *n.* (系统) 集成师

4.2.6 Budget

C-Registration Project—Budget for Releases 1 & 2			
Labor			
	Activities	Effort (PDs)	Cost
	Business Modeling	45	$31,500
	Requirements	90	$63,000
	Analysis & Design	130	$91,000
	Implementation	206	$144,200
	Test	140	$98,000
	Management	80	$56,000
	Deployment	50	$35,000
	Environment	90	$63,000
		TOTAL EFFORT: 831	
		TOTAL LABOR:	**$581,700**
NonLabor	(attached supporting details)		
	Travel and Accommodation:		0
	Freight & Duty:		0
	Services:		6,000
	Materials:		32,000
	Other Direct Charges:		7,500
		TOTAL NONLABOR:	**$45,500**
		TOTAL BUDGET:	**$627,200**

4.3 Iteration Plans

See the Course Registration System E1 Iteration Plan (11).

4.4 Project Monitoring and Control

4.4.1 Requirements Management Plan

The requirements for this system are captured in the Vision (1) and Stakeholder Requests (3) documents.

4.4.2 Schedule Control Plan

The project manager maintains a summary schedule showing the expected date of each milestone, and is part of the Status Assessment Report, as described in the reporting plan. The Status Assessment Report is provided to the IT Executive, who may use this to set new priorities or to recommend corrective action.

The summary schedule is derived from a detailed schedule maintained by the team managers. The line items in the detailed schedule are work packages assigned to individuals. Each individual who is assigned a work package provides completion information to his/her team manager on a weekly basis.

4.4.3 Budget Control Plan

Expenses are monitored by the project manager, and reported and assessed via the Status Assessment Report.

4.4.4 Quality Control Plan

All deliverables are required to go through the appropriate review process. The review is required

> **Glossary**
>
> capture *v.* 记录
> via *prep.* 经过，通过
> appropriate *adj.* 适当的

to ensure that each deliverable is of acceptable quality, using guidelines described in the Rational Unified Process (6) review guidelines and checklists.

In addition, defects will be recorded and tracked, and defect metrics will be gathered as described on the Zhonghua Software Process Website (10).

4.4.5 Reporting Plan

The Status Assessment report will be prepared by the Project Manager at least once per month. This includes:

- updated cost and schedule estimates
- summary of metrics

4.4.6 Measurement Plan

Standard project metrics, as described in the Zhonghua University Software Process Website (10), will be gathered and included in the Status Assessment Report.

4.5 Risk Management Plan

See Course Registration System Risk List (8).

4.6 Close-out Plan

The schedule will show a gradual roll-off of staff onto other projects. At least one developer will be retained part time by the IT Department after delivery to provide maintenance of the system.

A Post-mortem Report will be submitted to the IT Executive summarizing lessons learned, including an assessment of actual cost and schedule vs. predicted.

Glossary

checklist *n.* 一览表
metric *n.* 度量，规格
gradual *adj.* 逐渐的，逐步的
roll-off *n.* 转出，调走
retain *v.* 保留
post-mortem *adj.* 事后的
executive *n.* 执行官

Unit 2　Formulating Plans

 Post-reading exercises

Exercise 1

The development of the Course Registration System will be conducted using a phased approach. How many phases are mentioned according to the project phase plan? What's the major milestone that marks the completion of each phase?

Exercise 2

Please fill in the table according to Section 4 in the document.

	Content List
Release 1	
Release 2	

Exercise 3

Translate the following sentences into Chinese.

a) The objective of this Software Development Plan is to define the development activities in terms of the phases and iterations required for implementing a computerized class registration system for Zhonghua University.

_____

_____

b) This Software Development Plan describes the overall plan to be used by General Computers Corporation for developing the Course Registration System for Zhonghua University.

_____

_____

c) Other artifacts will be produced, as described in the project development case, but are not intended to be delivered to the maintenance organization.

d) A Work Breakdown Structure is being prepared, and will be provided in the next version of this document.

e) Each phase consists of development iterations in which a subset of the system is developed.

II Discussing by e-mail
Writing

Ⓐ Writing meaningful subject lines

E-mails are now considered as an important means of communication. In the IT industry, where efficiency and brevity are appreciated, electronic communication has become increasingly popular because of its speed and broadcasting ability. Here is a series of guidelines to help you write effective professional e-mails.

Communicating Professionally and Effectively by E-mail

—Subject line

A subject line that clearly identifies the body of the e-mail aids the reader to pay heed to the new subject under consideration. The subject line should be brief, does not need to be a complete sentence, and should clue the reader to the contents of the message. Recipients always scan the subject line in order to decide whether to open, forward, archive, or trash a message. Remember your message is one of many in your recipient's mailbox.

Always include a subject line in your message.

Almost all mailers present you with the subject line when you browse your mailbox, and it's often the only clue the recipient has about the contents when selecting messages to read.

Try to use a subject that is meaningful to the recipient as well as yourself.

Crafting a meaningful subject line will prompt people to open your e-mails and act on them quickly. For instance, when you send an e-mail to a company requesting information about a product, it is best to mention the actual name of the

product, e.g. "Product A Information".

Consider some examples of subject line entries:

× **Subject: Important! Read Immediately!**

What is important to you may not be important to your reader. Rather than brashly announcing that the contents of your message are important, write an informative headline that actually communicates at least the core of what you feel is so important: "Attention: All Group Members May Have to Work Overtime for the Next Month."

× **Subject: Meeting.**

The purpose of this e-mail might be a routine request for a meeting, an announcement of a last-minute rescheduling, or a summary of something that has already happened. There's no way to know without opening the message, so this subject line is hardly useful.

? **Subject: Follow-up on Meeting.**

This is a marginal improvement—provided that the recipient recognizes your name and remembers why a follow-up is necessary.

√ **Subject: Do we need a larger room for meeting next Fri?**

Upon reading this revised, informative subject line, the recipient immediately starts thinking about the size of the room, not about whether it will be worth it to open the e-mail. You'll get a faster response if your recipient can tell from the subject line that it's a real message from a real person.

If you are replying to a message but are changing the subject of the conversation, change the subject line as well.

The subject is usually the easiest way to follow the thread of a conversation, so changing the conversation without changing the subject can be confusing and can make filing difficult. Or better still, you can start a new message altogether.

Unit 2 Formulating Plans

B Discussing by e-mail

Jacky Chen wrote Jane an e-mail to discuss a problem Jane had brought up previously. Jacky first commented on Jane's proposal, and then pointed out his doubts about it. Lastly, he expressed his own opinion about what the plan should be.

From: jackychen@generalcomputers.com
To: janebaker@generalcomputers.com
CC:
BCC:
Subject: Maybe we can have a better solution.
Date: Tue, Oct 26, 2004

Hi, Jane,

  How are you doing with your report about the phase plan?

  After your called yesterday, I thought a lot about the issue. I totally agree with your opinion that Release 1 and Release 2 must be designed separately. You brought up a very good point.

  However, it seems to me that you thought Release 2 could be designed only after Release 1 was finished. If that's what you meant, I would suggest a slight alteration that will be workable.

  In my view, if we draw a clear line between the functions of each of the two releases, it is probable that we begin designing Release 2, or at least a part of Release 2, immediately after we complete Release 1's design. In this way, we will have fewer iterations,

> **Glossary**
>
> separately *adv.* 分别地
> bring up 提出

and we can start to implement R2 right after R1 is completed.

Please feel free to respond to this proposal.

Regards,

Jacky

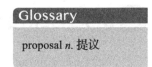

Glossary

proposal *n.* 提议

Now read this e-mail written by Jane to Mr. Zhang. This e-mail is a bit more formal than the previous one, as it is written from a subordinate to a superior. Compare the different styles.

From: janebaker@generalcomputers.com

To: davidzhang@generalcomputers.com

CC:

BCC:

Subject: Progress in the project plan.

Date: Wed, Oct 27, 2004

Attachments: phase_plan_status_report.pdf

Dear Mr. Zhang,

Good morning!

I've sent you a status report on the phase plan I am working on. The key part in this report is a review of the schedule for Release 2 design and implementation.

In our previous meetings, we did not separate the development schedules for Release 2 and Release 1. Tying their development together does not allow us to know what Release 2 should look like, without first completing Release 1.

Unit 2 Formulating Plans

Hence Jacky and I narrowed our possible solutions to two, as contained in the Mitigation Plans section of my report. One proposal calls for design work on Release 2 to proceed only after Release 1 is fully implemented. The other proposal raised by Jacky is that design work of Release 2 work in tandem with that of Release 1.

I tend to agree with Jacky's idea, but that may cause some confusion in our group. Maybe we can discuss this more fully at a later date.

Have a nice day!

Sincerely yours,

Jane Baker

Systems Analyst

General Computers Corporation

Glossary

hence *adv.* 因此，所以
narrow *v.* 使收缩，限制
mitigation *n.* 缓解，减轻
in tandem 一前一后地
tend *v.* 倾向于

Writing assignments

Imagine you are writing an e-mail to your partner to discuss an issue he/she raised. He/she thought the project schedule was a little too tight for the systems engineering group, and that it might be better to remove a week from the implementation phase and add it to the design & analysis phase. You want to tell him/her:

- You agree that the design & analysis phase should be given more time.
- However, as a member of the software engineering group, you are fully aware of the expected intensity during the implementation phase, and you contend that cutting out a week from that phase would be impossible.
- Maybe the schedule for the testing group is more flexible, and he/she may discuss with them about being allotted some of their time.

Write an e-mail to your superior, Mrs. Johnson Content to discuss the issue. When you write to a superior, use a more formal style. Include the following Content:

- Tell Mrs. Johnson that your group believes the budget for design & analysis is insufficient to accomplish its goals.
- Explain the reasons to her.
- Persuade her to contact the client in order to request increased funding.
- Send your e-mail to your partner as well as your teacher, eliciting their comments.

Try to improve your style and word choice well enough to persuade your superior.

C Guide to technical writing

No matter what your current or future job is, writing will be essential to your work because you will have to communicate your technical knowledge to others. Technical documents are a most frequently used type of writing in the IT industry, so training in technical writing will help ensure a smooth career path. Here is a series of guidelines to help you write successful technical documents.

Guidelines to Successful Technical Writing

—Outlining

When you write a technical document, not only must you decide what to include and what to exclude, you must also find a good way to arrange the information. Creating an outline can very effectively help us bring coherence to large amounts of information.

Formats may vary according to your needs and preferences and the body of information. For most short reports, a list of words or phrases may suffice. A more complex report, plan or proposal may require a systematic arrangement of topics, with a formal numbering and lettering system. In any case, never write the final draft of your report—long or short—without following some sort of outline.

Why may a writing assignment appear to fall easily into place for one writer, yet not another? One answer could well be that the successful writer devotes sufficient attention to planning and organizing before writing. A good working outline serves you in at least four important ways:

- It shows you which areas of information are to be investigated and gathered.
- It shows you which areas you can safely ignore.
- It enables you to schedule your work into manageable units of time.
- It gives you a global view of your report project, an overall sense of the contents, parts and organization of the report.

An *informal outline* is a working draft and includes a listing of main topics in the order the writer expects to present them. It's probably all you need for a short routine report.

In developing an informal outline, remember these following guidelines:

1. *List all the relevant topics in any order at first.* You can list the main topics to be covered in the order in which you find them discussed in company files.

2. *Identify the major groups of related information from your list.* For example, if you are writing a project plan, and you find that most of the information in your list of items pertains to project schedules, budgets, and quality control plans, you can cluster your information into categories under these topic headings.

3. *Arrange the information categories in an order that will best serve the readers' need to know.* The project schedule should appear first in the outline because scheduling information is important in all phases of the project. Then you can create budget categories from most common to least common. Thus, readers can quickly encounter the information they need.

A *formal outline* is a more detailed and systematic arrangement, using a special numbering system with subtopics under each major heading. There are two kinds of formal outlines: topic outlines and sentence outlines. A formal topic outline lists all the major topics and all the subtopics in a document by key words, and a formal sentence outline develops a topic outline by stating each point in a full sentence.

When you develop a formal outline from an informal or working outline, keep in mind these guidelines:

1. *Make the list of major topics broad enough to encompass your subject.* For example, the outline for a feasibility report would not be adequate if it did not include an evaluation of the existing system. Likewise, a project plan will not be complete without a "Assumptions and Constraints" part.

2. *Make the outline specific enough so you can discuss each topic in detail.* Thus, you can partition a topic into subtopics and, even further, sub-subtopics. This breakdown helps the reader understand each step or each aspect of the topic.

3. *Use logical notation and consistent formatting.* Notation is the system of numbers, letters, and other symbols marking the logical divisions of your outline. Format, on the other hand, is the arrangement, or layout, of your material. Proper notation and format show the subordination of some parts of your topic to others. Because your formal outline contains both major and minor parts, be sure that all sections and subsections are ordered, capitalized, lettered, numbered, punctuated, and indented to show how each part relates to other parts, and to the overall topic to be discussed.

4. *Make all items of equal importance parallel, or equal, in grammatical form.* Then your outline will emphasize the logical connections among related ideas.

5. *Present the topics in a logical sequence.* All information should appear in a logical order or sequence. Possibilities include chronological sequencing, spatial sequencing, for/against reasoning, problem cause/solution arrangement, cause/effect, simple to complex, etc.

6. *Be sure that all items in your outline add to the objectives of the document.* Omit irrelevant materials.

Unit 2 Formulating Plans

D Technical writing exercises

Exercise 1

Locate a short article (1,000 words maximum) from a journal in your field and make a topic or sentence outline of the article. Does the article conform to the principles of outlining discussed above? If not, suggest improvements. Discuss your conclusion in class or in one or two paragraphs.

Exercise 2

For each of the following report topics, select the most appropriate sequence to organize the subject.

Possible Sequences: *chronological sequencing, spatial sequencing, for/against reasoning, problem/causes/solution, cause/effect, comparison/contrast, simple to complex, sequence of priorities.*

a) a proposal for a micro-computer lab at your college

   spatial sequence (in which the outline follows the physical arrangement of parts)

b) a report describing your progress in developing a billing system for a course registration system

c) a report analyzing the desirability of a proposed R&D center in a new area

d) a proposal for a no-grade policy at your college

e) a report on any highly technical subject, written for a general reader

Exercise 3

Assume you are preparing a part of the software development plan for a public information system. The subject of your part is the managerial process. After group discussions and researching, you settle on the five following major topics:

- *staffing plan*
- *monitoring and controlling mechanics*

- *management objectives and priorities*
- *risk management*
- *assumptions, dependencies, and constraints*

Arrange these topics in the most sensible sequence.

When your topics are arranged, assume that subsequent discussions and further brainstorming produced the following list of subtopics:

- *kind of status reporting*
- *relative priorities among functionality, schedule, and budget*
- *minor risks*
- *reporting structure and frequency*
- *numbers and types of personnel*
- *time limit*
- *training programs*
- *the assumption that the information machines work as expected*
- *major risks*
- *budget limit*
- *project tracking system*
- *report contents/formats*
- *management philosophy*
- *contingency plans*
- *audit mechanisms*
- *required skill levels, start times, and durations for the personnel*

Arrange each subtopic (and perhaps some sub-subtopics) under its appropriate topic headings. Use an effective notation system and a good format to create the body of a formal sentence outline.

Unit 3

Specifying Customer Needs

明确需求

I   Software requirements specifications

II   E-mail & technical writing

Software requirements specifications
Reading

A An overview

Software Requirements Specification (SRS)

An SRS is a document describing the requirements of a computer system from the user's point of view. An SRS document specifies:

- the required behavior of a system in terms of input data, required processing, output data, operational scenarios and interfaces;
- the attributes of a system including performance, security, maintainability, reliability, verifiability, availability and safety requirements and design constraints.

Software Requirements Specifications provide us a good basis upon which we can both define a great specification and help us identify deficiencies in our past efforts.

The IEEE 830 Standard defines the benefits of a good SRS:

Establish the basis for agreement between the customers and the suppliers on what the software product is to do. The complete description of the functions to be performed by the software specified in the SRS will assist the potential users to determine if the software specified meets their needs or how the software must be modified to meet their needs. We use it as the basis of our contract with our clients all the time.

Reduce the development effort. The preparation of the SRS forces the various concerned groups in the customer's organization to consider rigorously all of the requirements before design begins and reduces later redesign, recoding, and retesting. Careful review of the requirements in the SRS can reveal omissions, misunderstandings, and inconsistencies early in the development cycle when these

Unit 3 Specifying Customer Needs

problems are easier to correct.

Provide a basis for estimating costs and schedules. The description of the product to be developed as given in the SRS is a realistic basis for estimating project costs and can be used to obtain approval for bids or price estimates. In reality, we do use the SRS as the basis for our fixed price estimates.

Provide a baseline for validation and verification. Organizations can develop their validation and verification plans much more productively from a good SRS. As a part of the development contract, the SRS provides a baseline against which compliance can be measured. Actually, we use the SRS to create the Test Plan.

Facilitate transfer. The SRS makes it easier to transfer the software product to new users or new machines. Customers thus find it easier to transfer the software to other parts of their organization, and suppliers find it easier to transfer it to new customers.

Serve as a basis for enhancement. Because the SRS discusses the product but not the project that developed it, the SRS serves as a basis for later enhancement of the finished product. The SRS may need to be altered, but it does provide a foundation for continued production evaluation. This is often a major pitfall when the SRS is not continually updated with changes.

B A sample of software requirements specifi-cation

Read the following software requirements specification. This is only a preliminary document, but it sheds much light on what a software requirement specification is. For the first time, please only scan the whole document. Keep these questions in mind and try to answer them after scanning. Time limit: 10 minutes.

- What does LCS do?
- What protocols are to be used in this project?
- What kind of database is used in this project?

Title:	Specification of VTS	Date:	6/27/20xx
Project:	Vehicle Tracking System	Author:	Jacky Chen
Request for Comments		Start Time:	
		Project Manager:	David Zhang

Glossary

vehicle *n.* 车辆

track *v.* 跟踪

memorandum *n.* 备忘录

revision *n.* 修订, 修正

memo *abbr.* =memorandum

handler *n.* 处理程序

INTERNAL MEMORANDUM

Requirement Specifications of VTS for General Computers Corporation

Revision

Version	Author	Time	Memo
0.9	Jacky Chen	20xx-09-06	
0.9.1	Jacky Chen	20xx-09-08	

Contents

1. **General**

2. **Definitions and Abbreviation**

3. **Overview**

4. **Function**

   4.1 Function List

      4.1.1 TCP Server Handler

      4.1.2 InQueue

      4.1.3 OutQueue

      4.1.4 G-Tech Server Request Handler

Unit 3 Specifying Customer Needs

4.1.5 LCS Command Handler

4.2 Data Flow

4.3 Configuration

5. Interfaces

5.1 Communication Protocol

5.2 Interfaces between VTS and G-Tech Server

5.3 Interfaces between VTS and LCS

6. Technical Requirement

6.1 Stress Load

6.2 Database

6.3 Language

6.4 Application Server

> **Glossary**
>
> device *n.* 装置, 设备
> architecture *n.* 体系结构, 架构
> forward *v.* 发送

1. **General**

This is the specification of VTS. This document describes the functions, technical requirements and interfaces of VTS. VTS is a vehicle location tracking system based on GPRS network.

The Vehicle Tracking Communication Software will provide communication between the G-Tech vehicle tracking <u>devices</u> (GPRS+GPS) and vehicle tracking server. Under the current system <u>architecture</u>, the G-Tech devices communicate through the GPRS network to a server at G-Tech. G-Tech then <u>forwards</u> the location data to a server at GCC Computers running our vehicle tracking application.

The new Vehicle Tracking Communication Software will eliminate the need for the G-Tech devices to communicate to the G-Tech server and will interface directly to the GCC Computers vehicle tracking server. It will allow the G-Tech devices to establish a socket connection, receive the incoming location data, queue the data, send a reverse geocode request to the GCC Computers G-Tech server, incorporate the reverse geocoded address into the location record, queue the geocoded records and pass them to the existing vehicle tracking server software. The existing software will write the records to the existing SQL database and provide the user interface. The existing software will also initiate a message to the G-Tech units. The new Vehicle Tracking Communication Software must be able to receive the messages, forward them to the G-Tech devices through the open socket connection, receive the <ACK> response from the G-Tech units and provide an "OK" or "FAIL" response to the existing server software.

Glossary

socket *n.* 套接口
queue *v.* 使排入队伍
reverse *adj.* 反向的
geocode *n.* =geographical code 地理代码
existing *adj.* 现有的
initiate *v.* 开始, 启动
extensible *adj.* 可扩展的

2. Definitions and Abbreviation

Abbreviations	Descriptions
VT	Vehicle Tracker
VTS	Vehicle Tracking System
LCS	Location Collector System
XML	Extensible Markup Language
JMS	Java Message Service
GPRS	General Packet Radio Service

Unit 3 Specifying Customer Needs

(Continued)

Abbreviations	Descriptions
TCP	Transmission Control Protocol
TTT	Text-to-Text
Geofence	By setting up a geofence, a dispatcher can set up a geographic boundary for a particular vehicle.
IMEI	International Mobile Equipment Identity

Glossary

protocol *n.* 协议
geofence *n.* 地理栅栏
dispatcher *n.* 调度程序
boundary *n.* 边界, 分界线
identity *n.* 身份, 特性
integrate *v.* 结合, 合并
asset *n.* 资产
assist *v.* 帮助, 辅助
consumer *n.* 消费者
empower *v.* 授权, 准许
investment *n.* 投资
third-party *adj.* 第三方的

3. Overview

Tracking system is used to track VT location. VT will send location record to server and server will send commands to VT when in need. The whole tracking system is integrated by VT, VTS, LCS and G-Tech Server.

VT is a tracking device installed in the vehicle. VT enables you to locate, track and monitor your mobile assets throughout North America. For businesses, the tools assist with the complex task of managing mobile assets. Consumers are empowered with the ability to keep watch of their investments from any computer connected to the Internet.

VTS is the communication server and location process server. VTS gets location record sent from VT, and can watch 1,000 vehicles in the same time. The protocol is handled in VTS.

LCS is a location collector database with TCP server and command server. It will send command to VTS when in need.

G-Tech Server is a third-party system, which will provide address info based on the location from

VTS and return this address info in XML format to VTS;

From the following picture of tracking system design architecture, VTS is the core of the whole system.

Glossary

authentication *n.* 鉴别，确认

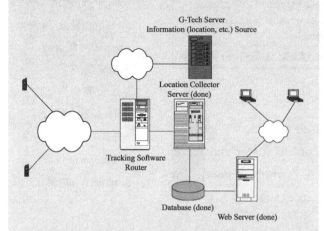

The VTS is a piece of tracking software which will be used to track vehicle moving. A VT, which is installed in a vehicle, will login into VTS and send location info to VTS. After it gets location from VT, VTS will get address info from G-Tech Server which is a third party system. VTS also receives commands from LCS, and will then send these commands to the particular VT.

4. Function

4.1 Function List

VTS is designed for providing the following functions:

- VT device management (Authentication, Connection and Charging Info);

- VT protocol encoding and decoding;
- getting location record from VT by opened TTT connection;
- providing VT location info to LCS based on TTT protocol, getting VT commands(C0, C1, C2) from LCS, and sending these commands to VT;
- receiving third-party information by VT location and returning the acquired information from third-party information to LCS;
- Management of TTT connection between VT and VTS, VT connectivity detection.

VTS contains the following modules: TCP Server Handler, InQueue, OutQueue, G-Tech Server Request Handler and LCS Command Handler.

4.1.1 TCP Server Handler

This component will open and maintain a socket for direct communication with a G-Tech device when a G-Tech device attempts TCP communications with the specified IP address/port. It must verify that the records are in the correct format and that the IMEI is a valid IMEI in the vehicle tracking database. If the IMEI does not exist in the database, it must not accept data from the device and should close the socket connection. It should send an error message to the existing server that the IMEI attempted to log in but it was not a

Glossary

encode *v.* 编码
decode *v.* 解码, 译解
acquire *v.* 获取
connectivity *n.* 连通性
component *n.* 组件
attempt *v.* 试图, 尝试
verify *v.* 验证, 校验

valid IMEI in the vehicle tracking database. If the records are not in the correct format, it should reject the records and close the connection. It should send an error message to the existing server that the IMEI attempted to send records but they were not in the correct message format.

4.1.2 InQueue

This component will queue incoming messages and pass them to the G-Tech Server Request Handler. Its main purpose is to manage the incoming message flow to ensure that incoming messages are not lost.

4.1.3 OutQueue

This component will queue incoming messages from the G-Tech Server Request Handler and pass them to the existing vehicle tracking software. Its main purpose is to manage the message flow to ensure that messages are not lost.

4.1.4 G-Tech Server Request Handler

This component will receive incoming messages from the inQueue. If the incoming message contains a latitude/longitude, it will use the latitude/longitude to send a reverse geocode request to the existing G-Tech engine. It will receive the reverse geocoded address from the G-Tech engine and insert it into the message. It will then forward the modified message to the outQueue.

If the incoming message does not contain a latitude/longitude, it will simply forward the message to the outQueue.

Glossary

latitude *n.* 纬度

longitude *n.* 经度

4.1.5 LCS Command Handler

This component will receive reverse commands which come from LCS, and these commands will be put into the outQueue.

4.2 Data Flow

A communication loop step by step:

1) VTS is started and it starts TCP listener at the specified IP and Port, and loads all VT IMEI from database.

2) The VT is powered on and comes into GPRS coverage areas, and then the VT attaches to the GPRS network and creates an active GPRS connection.

3) The VT creates a TCP connection to the fixed IP and Port, and VTS gets the connection request at the listener and accepts the connection.

4) The VT sends login message, when the IMEI of the VT is sent as the id. VTS verifies IMEI and decides to accept or reject this connection.

5) Connection is set up, and VT sends Location Record in T-TCP protocol to the VTS.

6) The VTS gets Location Record, sends location to G-Tech Server by GTHttpRequest class, and gets return in XML format.

7) The VTS processes the returned address info from G-Tech Server and sends address info to the LCS.

8) If the LCS needs to send commands to the

Glossary

coverage *n.* 覆盖率; 有效区域

VTS, the VTS will encode commands and send encoded commands to the particular VT.

4.3 Configuration

Configuration of VTS is saved in an XML file. All parameters in the configuration will be used in system.

VTS-VT Port, the listener port of VTS.

G-Tech IP, a fixed IP address from which VTS can get address info.

VTS-LCS port, used to listen LCS connection request.

LCS IP and Port, configured for communication between VTS and LCS.

Log file name, used to save runtime info.

Database JDBC env string, used to set up JDBC connection.

5. Interfaces

5.1 Communication Protocol

G-Tech VT T-TCP protocol will be used in VTS to communicate with VT and LCS.

IMEI of GPRS device will be used as id for authentication. When a VT logs in, it will send its own IMEI to VTS, and VTS will verify this id. Only legal device can log in. A 3-byte appid will be used as *application identity*. This appid also will be verified by VTS. When VTS sends commands to VT, VT will check the appid to make sure that only legal sources can do that.

Glossary

configuration *n.* 配置, 结构
parameter *n.* 参数, 变量
log *n.* (运行) 记录, 日志

5.2 Interfaces between VTS and G-Tech Server

The connection between VTS and G-Tech Server is HTTP-based. A GTHttpRequest class will be provided by GCC to communicate with G-Tech Server. The interfaces will send location records to G-Tech Server by VTS. And G-Tech Server will return address info in XML format.

The following are the interfaces to G-Tech Server:

GTHttpRequest.sendLocation()

GTHttpRequest.getAddress()

5.3 Interfaces between VTS and LCS

The interface between VTS and LCS is TCP-based. VTS and LCS work peer to peer, both can listen to and close the connection.

L0 : not found in protocol document

C0 : OTA command

M0 : not found in the protocol document

6. Technical Requirement

6.1 Stress Load

Min 1000 T-TCP connections

6.2 Database

SQL Server 20xx

6.3 Language

Java (JDK1.4)

6.4 Application Server

Jboss

Post-reading exercises

Exercise 1

A company has a good development environment, including Min 1000 T-TCP connections, SQL Server 20xx beta, Java (JDK 1.3) and JBoss as application server. Does this development environment meet those technical requirements? If not, please state why.

Exercise 2

Fill in the blanks with proper descriptions or abbreviations.

Abbreviations	Descriptions
VT	
VTS	
	Location Collector System
	Section Locator
XML	
	Java Message Service
GPRS	
	Transmission Control Protocol
	By setting up a geofence, a dispatcher can set up a geographic boundary for a particular vehicle.
IMEI	

Exercise 3

Translate the following passage.

A communication loop step by step:

1) VTS is started and it starts TCP listener at the specified IP and Port, and loads all VT IMEI from database.

Unit 3 Specifying Customer Needs

2) The VT is powered on and comes into GPRS coverage areas, and then the VT attaches to the GPRS network and creates an active GPRS connection.

3) The VT creates a TCP connection to the fixed IP and Port, and VTS gets the connection request at the listener and accepts the connection.

4) The VT sends login message, when the IMEI of the VT is sent as the id, VTS verifies IMEI and decides to accept or reject this connection.

5) Connection is set up, and VT sends Location Record in T-TCP protocol to the VTS.

II E-mail & technical writing
Writing

A Salutation, greeting & identification

E-mails are now considered an important means of communication. In the IT industry, where efficiency and brevity are appreciated, electronic communication has become increasingly popular because of its speed and broadcasting ability. Here is a series of guidelines to help you write effective professional e-mails.

Communicating Professionally and Effectively by E-mail

— Opening an e-mail

Salutation

√ **Always be careful about salutations.**

Salutations are tricky, especially if you are e-mailing cross culturally. Frequently, titles are different for men and women, and you may not be able to tell which you are addressing. Also the family name is first in some cultures and last in others. Finally, honorific titles may vary depending on status or age. So don't feel bad if you have trouble figuring out which salutation to use: it is a difficult problem. Universally, however, consider investing time to get more information about addressees in other cultures.

√ **Use titles only when you are sure.**

When you are writing to English native speakers, it is a bad idea to use "Sir" or "Mr." unless you are absolutely certain that your correspondent is male. Similarly, it is probably safer to use "Ms." instead of "Miss" or "Mrs." unless you know the preference of the woman in question.

Unit 3 Specifying Customer Needs

√ **Use familiar and simple salutations where it is appropriate.**

If you are corresponding with an American, or a very familiar friend from other English speaking countries, using the first name is usually appropriate. Thus, you can usually get away with a "Dear" and the first name. For example:

Dear Chris,

Here you are covered regardless of whether Chris is male or female. Beware of using a diminutive if you aren't certain your correspondent uses it. It might irritate Judith to be called Judy; Robert might hate being called Bob.

If you are addressing a group of people, you can say "Dear" plus the collective name. For example:

Dear Project Managers,

Or:

Dear Los Angeles Lakers Fans,

Americans often use a simple "Hi" before people's first name:

Hi, Charles,

× **Use simple salutations for people with formal traditions.**

Again, you must be very careful about cultural differences. The East Coast of the United States is more formal than the West Coast. Germans are even more formal: they can work side-by-side for years and never get around to a first-name basis. Starting a message to Germany with *Dear Hans* might be a bad idea.

Greeting & Identification

√ **Initiate e-mails with greetings only when you have to.**

To greet or not to greet? Although greetings can help start things off on a friendlier or a more formal note, many longtime Web users eschew greetings in e-mail, preferring to launch into their message without much ado. It's clean, neat, simple, and appropriate for almost every occasion. Consider joining this camp. It sure makes life easier.

"Good Morning!" and "Good Afternoon!" don't make a lot of sense with

e-mail, as the earth may have rotated by the time the recipient gets around to reading it. "Good Day!" may sound stiff to American ears, though it may be common in parts of the former British Empire.

√ **Understand the importance of identifying yourself.**

When a person receives an e-mail from someone he doesn't know, he/she cares more about that fact than about how he/she is addressed. Therefore, before you send an e-mail, particularly to someone who doesn't know you, it might be better to find answers to the following questions:

- How did you learn of your crrespondent?
- What do you want from your correspondent?
- Who are you?
- Why should your correspondent pay attention to you? (If you can't find a satisfactory answer to this question, you probably should not include him/her on your mailing list.)

√ **Put your identity at the beginning of an e-mail.**

Including identity information in a signature is better than not providing it, but inserting it at the top is preferable for the following reasons:

- If problems with the transmission of the e-mail arise, the end is more likely to be lost than the beginning.
- A lot of people get more than twenty messages per day, and process them quickly. If you don't establish early on who you are, the receiver may well delete your message without a full reading.
- Your identity frames the context of your message.

Please take note of the following example.

Unit 3 Specifying Customer Needs

Dear Mr. Sherwood:

I am an engineer at General Computers Corporation. I sat next to your colleague, Mr. Wang, in an international conference in Bonn last week, and he mentioned that you are interested in writing a book about software project processes. I have a similar interest, and would be very pleased if you can take a look at the outline I'm sending to you.

B E-mails about software functions

Imagine you are writing an e-mail to your partner to discuss an issue he/she raised. He/she thought the vector-based image manipulation software, which your team is currently developing, should also be able to edit bitmap images. You want to tell him/her:

- You appreciate his/her effort to expand the functionality of the software.
- However, it may make things too complicated, and hinder the current progress.
- Maybe it's better to add limited bitmap-based functions to the software, but not too much.

Send this e-mail to your partner as well as your instructor, and ask them for their feedback.

C Guide to technical writing

No matter what your current or future job is, writing will be essential to your work because you will have to communicate your technical knowledge to others. Technical documents are a most frequently used type of writing in the IT industry, so training in technical writing will help ensure a smooth career path. Here is a series of guidelines to help you write successful technical documents.

Guidelines to Successful Technical Writing
—Developing good paragraphs

In the previous unit, we learned how to organize information in outline form. The reader also needs the content of the document to be accessible. Any reader approaches a message with expectations about organization as well as content.

Thus, most useful messages—whether in the form of a book, report, proposal, news article, letter, or memo—usually adhere to a common organizing pattern:

INTRODUCTION
BODY
CONCLUSION

The *introduction* previews the discussion, revealing the subject and purpose of the message. The *body* elaborates on the idea which is stated or implied in the introduction, with a full explanation of the main point. The *conclusion* naturally terminates the message in summary form.

The building blocks of all technical writing are paragraphs. The standard paragraph is a group of sentences focusing on one main organizing point. That main organizing point, is called the *topic sentence*. Most paragraphs in technical writing follow an introduction-body-conclusion format. The key to writing good paragraphs is to begin with a clear topic (orienting) sentence that states a generalization. When writing a topic sentence, remember to follow these guidelines:

1. *Include a topic sentence in every paragraph.* Readers always look to the first one or two sentences in a paragraph to orient themselves. The topic sentence ties all other sentences in the paragraph together to convey a single message. Without it the reader cannot possibly grasp your exact meaning.

2. *Before you can write a good topic sentence, identify your purpose, based upon your knowledge of your readers' needs.* Only in this way can you tailor your

Unit 3 Specifying Customer Needs

topic sentence to meet most needs.

3. *The topic sentence must be focused enough to be covered in one paragraph.* You must think of ways to make the topic sentence more informative so it expresses your point most accurately. Avoid broad or abstract topic sentences.

In addition to a topic sentence, ***paragraphs include sentences with details that confirm the topic sentence.*** In constructing those sentences include information the reader requires in order to understand the paragraph's main idea. Supporting sentences may include:

1) examples of the topic;

2) facts, statistics, evidence, details, or precedents that confirm the topic;

3) quotes, paraphrases, or summaries of evidence on the topic;

4) descriptions of events that relate to the topic;

5) definitions of terms connected with the topic;

6) explanations of how something works;

7) descriptions of the physical appearance of pertinent objects, areas, or people.

The following patterns for presenting information can be effective ways to organize paragraphs or entire sections of a document. Select the device that will best aid your readers in understanding and using the information.

1) Ascending/descending order of importance pattern: this information sequencing device clearly prioritizes information for the reader.

2) Cause/effect pattern: shows the relationship between events.

3) Chronological pattern: material is presented in its order of occurrence.

4) Classification pattern: involves grouping items in terms of certain characteristics and shows your readers the similarities within each group.

5) Partitioning pattern: separates a topic or system into its individual features.

6) Comparison/contrast pattern: focuses on the similarities or differences between subjects.

7) Defining pattern: explains the meaning of a term that refers to a concept,

process, or object.

8) Spatial referencing pattern: groups information according to the physical arrangement of the subject.

Your concluding statement in the paragraph ends the discussion of the central idea. It usually ties the paragraph together by summarizing, interpreting, or judging the facts. If the paragraph is part of a longer report, your last sentence should help the reader make the transition to the next.

Finally, the introduction-body-conclusion structure should serve most of your needs in drafting paragraphs for technical writing. Begin each paragraph with a good topic sentence and you will stay on target.

D Technical writing exercises

Exercise 1

Select three typical paragraphs from any technical document you may have written, and examine the structure of the paragraphs. Does each contain a topic sentence? Are the topic sentences fully developed? Do these paragraphs contain concluding sentences? Discuss these questions with your partner, and work together to improve the paragraphs according to the guidelines presented above.

Exercise 2

The following topic sentences either provide no direction, are unfocused, or are not sufficiently informative. Revise them.

Examples:

1) Writing is a complex skill. (unfocused)

   Revised: Writing is a process that involves a series of deliberate decisions about audience needs, purpose, content, arrangement, and style.

2) A Mercedes-Benz is a great all-purpose car. (not sufficiently informative)

Unit 3 Specifying Customer Needs

Revised: A Mercedes-Benz offers safety, performance, durability, and luxury.

a) Technical writing is a popular profession.

b) Computer software has changed radically in the last decade.

c) Our team consists of more men than women.

d) This new module is great.

e) The name of the project is Acme Accounting System.

Exercise 3

Rewrite Section 4 from the VTS Requirements Specifications in Part 3.3 according to the template below. This template provides sentence patterns of typical topic sentences in respective subsections. Develop good sentences around these topic sentences.

Template:

1) Overview

The software is designed for providing the following functions:

<*Summarize the major functions the product must perform or must allow the user to perform. A high level summary (such as a bullet list) is needed here. Organize the functions to make them understandable to any reader of the SRS. A picture of the major groups of related requirements and how they relate is often effective, such as a top level data flow diagram or object class diagram.*>

allow/realize/require/use/have/...

2) Features

<Module 1>

2.1.1 Description and Priority

<*Provide a short description of the feature and indicate whether it is of high, medium, or low priority.*>

2.1.2 Response Sequences

<List the sequences of user actions and system responses that stimulate the behavior defined for this feature. These will correspond to the dialogue elements associated with use cases.>

2.1.3 Functional Requirements

<Itemize the detailed functional requirements associated with this feature. Include how the product responds to anticipated error conditions or invalid inputs.>

<Module 2>

2.2.1 Description and Priority

2.2.2 Response Sequences

2.2.3 Functional Requirements

...

Unit 4

Describing Designs

设计简述

I   Software design documents

II   E-mail & technical writing

Software design documents
Reading

A An overview

Software Design Specification

This document contains the complete design description of the System. This includes the architectural features of the system down through details of what operations each code module will perform and the database layout. It also shows how the use cases promised in the SRS will be implemented in the system using this design.

The primary audience of this document is the software developer.

- It may serve as training materials for new project members, imparting to them enough information and understanding about the project implementation, so that they are able to understand what is being said in design meetings, and won't feel as if they are drowning when they are first asked to create or modify source code.

- It may serve as "objective evidence" that the designers and/or implementers are following through on their commitment to implement the functionality described in the requirements specification.

B A sample of software design document

Read the following software design document. For the first time, please only scan the whole document. Keep these questions in mind and try to answer them after scanning. Time limit: 10 minutes.

Unit 4 Describing Designs

- What does this design document mainly describe?
- Which communications technology is VTS based on?
- How many functions does VTS provide?

Title:	Design of VTS	Date:	06/30/2005
Project:	Vehicle Tracking System	Author:	Jacky Chen; Sharon Wang
Request for Comments		Start Time:	01/18/20xx
		Project Manager:	David Zhang

> **Glossary**
>
> modify *v.* 修改
>
> multi-socket *adj.* 多套接口的

INTERNAL MEMORANDUM

Vehicle Tracking System Design Document for General Computers

Version 1.3

Revision

Version	Author	Time	Memo
0.5	Sandy Davis	20xx-01-21	First version.
0.9	Sandy Davis	20xx-02-12	
1.0	Jacky Chen, Sandy Davis	20xx-02-15	Modify after multi-socket connection test (100 over one thread).
1.1	Jacky Chen	20xx-03-19	LCS access and IMEI authentication modified.
1.2	Jacky Chen	20xx-05-20	LCS–VTS two connection modes added.

91

(Continued)

Version	Author	Time	Memo
1.3	Jacky Chen	20xx-06-30	LCS–VTS protocol finalized.

Glossary

finalize *v.* 完成，定案

Contents

1. General

2. Abbreviation and Definition

3. Overview

   3.1 VTS Functions

      3.1.1 TCP Server Handler

      3.1.2 InQueue

      3.1.3 OutQueue

      3.1.4 GT Request Handler

      3.1.5 LCS Command Handler

   3.2 VT-VTS Communication Flow

   3.3 Configuration

4. System Architecture

   4.1 Communication Agent

   4.2 Application Server

5. Communication Agent

   5.1 Overview

   5.2 Modules

      5.2.1 Connection Manager

Unit 4 Describing Designs

5.2.2 Session Manager

5.2.3 IO Processor

5.2.4 JMS inQueue

5.2.5 LCS inQueue

5.2.6 OutQueue

5.2.7 JMS Message Processor

5.2.8 LCS Message Processor

5.3 Data Flow

5.4 Protocol

    5.4.1 VTS/VT Protocol

        5.4.1.1 Interactions

        5.4.1.2 Protocol format

            5.4.1.2.1 Login Format

            5.4.1.2.2 Message Format

            5.4.1.2.3 Acknowledge Format

    5.4.2 VTS/LCS Protocol

        5.4.2.1 VTS to LCS

            5.4.2.1.1 OTA Command

            5.4.2.1.2 Message Record

            5.4.2.1.3 Location Record

        5.4.2.2 LCS to VTS

            5.4.2.2.1 OTA Command

            5.4.2.2.2 Message Record

    5.4.3 VTS/GT Protocol

6. Application Server

  6.1 Overview

> **Glossary**
>
> session *n.* 对话（期），会话
>
> IO = input/output 输入输出
>
> acknowledge *v.* 肯定响应

6.2 Architecture

6.2.1 JMS Manager

6.2.2 MDB

6.2.3 GT Access Bean

6.2.4 LCS Access Bean

1. General

This design document describes the software architecture, module design, data flow and function interfaces of VTS (Vehicle Tracking System).

VTS is a vehicle location tracking system based on GPRS network. The Vehicle Tracking & Communication Software will provide communication between the G-Tech vehicle tracking devices (GPRS+GPS) and vehicle tracking server. Under the current system architecture, the G-Tech devices communicate through the GPRS network to a server at G-Tech. G-Tech then forwards the location data to a server at GCC running our vehicle tracking application.

The new Vehicle Tracking & Communication Software will eliminate the need for the G-Tech devices to communicate to the G-Tech server and will interface directly to the GCC vehicle tracking server. It will allow the G-Tech devices to establish a socket connection, receive the incoming location data, queue the data, send a reverse geocode request to the GCC server, incorporate the reverse geocoded address into the location record, queue the geocoded records and pass them to the existing vehicle tracking server software. The existing software will write the records to the existing SQL database and provide the user interface. The existing software will also initiate a message to the G-Tech units. The new Vehicle Tracking & Communication Software must be able to receive the messages, forward them to the G-Tech devices through the open socket connection, receive the <ack> response from the G-Tech units and provide an "OK" or "Fail" response to the existing server software.

2. Abbreviation and Definition

Abbreviations	Descriptions
VTS	Vehicle Tracking System
VT	Vehicle Tracker
TCP	Transmission Control Protocol
T Protocol	T-acknowledged TCP-Based Protocol
J2EE	Java x Enterprise Edition
App Server	Application Server
EJB	Enterprise JavaBeans
JMS	Java Message Service
MDB	Message Driven Bean
MT	Mobile Terminal
GT	G-Tech Server
LCS	Location Collector Server
XML	Extensible Markup Language
Geofence	Geographic fence. By setting up a geofence, a dispatcher can set up a geographic boundary for a particular vehicle.
IMEI	International Mobile Equipment Identity
GPRS	General Packet Radio Service

3. Overview

Tracking system is used to track VT location. VT will send location record to server and server will send commands to VT when in need. The whole tracking system is integrated by VT, VTS, LCS and GT.

VT, vehicle tracker, is a tracking device installed in the vehicle. VT enables you to locate, track and monitor your mobile assets throughout North America. For businesses, our tools assist with the complex task of managing mobile assets. Consumers are empowered with the ability to keep watch of their investments from any computer connected to the Internet;

VTS, contains the Communication Agent and a location process server.

VTS gets location record sent from VT, and can watch 1000 vehicles at the same time. The protocol is handled in VTS;

LCS, is a location collector database with TCP server and command server. It will send command to VTS when in need;

G-Tech Server, is a third part system, which will provide address info based on the location from VTS and return this address info in XML format to VTS.

VTS is a tracking software which will be used to track vehicle moving. VT which is installed in vehicle will login into VTS, and send location info to VTS. After it gets location from VT, VTS will get address info from GT, which is a third party system. VTS will also get commands from LCS. These commands will be sent to particular VTs by VTS.

3.1 VTS Functions

VTS is designed for providing the following functions:

- SQL Server 20xx database access. It will get VT info from Database which will be used for authentication;
- VT protocol encode and decode;
- Get location record from VT by opened T-TCP connection;
- Provide VT location info to LCS based on T-TCP protocol; get VT commands (L0,C0,M0) from LCS, and send these commands to VT; Get third-party information by VT location and return the acquired information from third-party information to LCS;
- Management of T-TCP connection between VT and VTS, VT connectivity detection.

VTS contains the following modules: TCP Server Handler, InQueue, OutQueue, GT Request Handler and LCS Command Handler.

3.1.1 TCP Server Handler

This component will open and maintain a socket for direct communication with a G-Tech device when a G-Tech device attempts TCP connections with the specified IP address/port. It must verify that the records are in the correct format and that the IMEI is a valid IMEI in the vehicle-tracking database. If the IMEI does not exist in the database, it must not accept data from the device and should close the socket connection. It should send an error message to the existing server software that the IMEI attempted to log in but it was not a valid IMEI in the vehicle-tracking database. If the records are not in the correct format, it should reject the records and close the connection. It should send an error message to the existing server software that the IMEI attempted to send records but they were not in the correct message format.

3.1.2 InQueue

This component will queue incoming messages and pass them to the GT Request Handler. Its main purpose is to manage the incoming message flow to ensure that incoming messages are not lost.

3.1.3 OutQueue

This component will queue incoming messages from the GT Request Handler and pass them to the existing vehicle tracking software. Its main purpose is to manage the message flow to ensure that messages are not lost.

3.1.4 GT Request Handler

This component will receive incoming messages from the inQueue. If the incoming message contains a latitude/longitude, it will use the latitude/longitude to send a reverse geocode request to the existing GT engine. It will receive the reverse geocoded address from the GT engine and insert it into the message. It will then forward the modified message to the outQueue.

If the incoming message does not contain a latitude/longitude, it will simply forward the message to the outQueue.

3.1.5 LCS Command Handler

This component will receive reverse commands which come from LCS, and these commands will be put into the outQueue.

3.2 VT-VTS Communication Flow

A communication loop step by step:

1) VTS is started and it starts TCP listener at the specified IP and Port, and loads all appid from database;

2) VT is powered on. When it comes into GPRS coverage areas, VT attaches to the GPRS network and activates GPRS connection;

3) VT creates a TCP connection to the fixed IP and Port. VTS gets the connection request at the listener and accepts the connection;

4) VT sends login message. The IMEI of VT is sent as the id. VTS verifies IMEI and decides to accept or reject this connection;

5) A connection is set up. VT sends Location Record via T-TCP protocol to VTS;

6) VTS gets Location Record, sends location to GT by GTHttpRequest class, and gets return in XML format;

7) VTS processes address info returned from GT and sends address info to LCS;

8) If LCS needs to send command to VTS, VTS will encode command, and send encoded command to particular VT.

3.3 Configuration

The configuration of VTS is saved in an XML file. All parameters in the configuration will be used in the system.

VTS–VT Port, this is the listener port of VTS.

GT IP, a fixed IP address from which VTS can get address info.

VTS–LCS Port, used to listen LCS connection request.

LCS IP and Port, configured for communication between VTS and LCS.

Log File Name, used to save runtime info.

Database JDBC env. string, used to set up JDBC connection.

4. System Architecture

VTS is integrated by two independent sub-systems: communication sub-system and application sub-system. Communication sub-system will process the communication between VT and VTS, accept VT login and data, and forward the network message to Application sub-system via JMS or LCS. Application sub-system will process VT L0 message, send L0 message to GT server, and forward the GT returned result to LCS, at last Application sub-system will send LCS returned message to VT.

4.1 Communication Agent

Communication Agent is an independent application, and is able to run in an independent computer. It will listen at the specified port and process the T-TCP connection request, verify the VT validity, set up TCP connection with VT and exchange data. Communication Agent makes asynchronous message exchange with Application Server by JMS. While VT sends location record to Communication Agent, Communication Agent will forward location record (L0) to Application Server by JMS, and Application Server will return the message and processed result back to Communication Agent, and Communication Agent will send processed message to VT.

Communication Agent is based on JAVA (JDK 1.4), with new features like Asynchronous Socket. Asynchronous IO will be used to improve server performance.

> **Glossary**
>
> independent *adj.* 独立的
> sub-system 子系统
> asynchronous *adj.* 异步的
> JDK = Java Development Kit Java 开发工具包

4.2 Application Server

The Application Server is based on J2EE AppServer. It will take charge of message process which is sent from the Communication Agent. Application Server will use JMS to communicate with the Communication Agent, and it will use MDB for JMS. When a message is sent from Communication Agent to Application Server, it will be processed by MDB. J2EE Application Server can do dynamic MDB loading according to different situations, and it can balance connection loads according to different system loads. By these, Application Server can process mass parallel messages. MDB is the entry interface of message process. In MDB, Application Server will call EJB object to process specified logical transaction according to VT message type, such as GT server access, LCS server access, etc. Finally MDB will return processed message to Communication Agent by JMS.

Because of the deployment of JMS, MDB and EJB, Application Server can effectively process VT message, and have extra-strong scalability.

Glossary

dynamic *adj.* 动态的
balance *v.* 使……平衡
mass *adj.* 大量的, 海量的
parallel *adj.* 并行的
effectively *adv.* 有效地
scalability *n.* 可伸缩性

5. Communication Agent

5.1 Overview

Communication Agent will process TCP communication between VT and VTS, forward VT message to Application Server, receive VT update command from Application server and send VT update command to VT.

Communication Agent is based on JDK 1.4, based on asynchronous socket process. Communication Agent can provide over 1000 connection threads.

5.2 Modules

Communication Agent includes the following modules: connection manager, session manager, JMS inQueue, LCS inQueue, outQueue, IO processor, JMS message processor and LCS message processor.

When Communication Agent starts, it will invoke EJB to get all IMEI info in LCS database.

TCP connection request from VT to VTS is processed (accepted/rejected) by Connection Manager, which listens to specified ports. After a connection is accepted, login message will be sent by VT, and Session Manager will process login message to create a session. After a session is created, all data communications will be processed by IO processor.

5.2.1 Connection Manager

Connection Manager manages all VT TCP connections (accepts/closes connection), controls and maintains all TCP connection statuses. When system starts up, it will listen at the specified TCP port, when VT creates a TCP connection to this port, Communication Agent will accept the TCP connection. Connection Manager will provide interface to close/drop connection. If connection is invalid or should be closed, close interface of

Glossary

invoke *v.* 调用，启用

invalid *adj.* 无效的，非法的

Connection Manager will be invoked to close socket connection.

5.2.2 Session Manager

Session Manager is a working thread which manages all VT-VTS TCP communication sessions. It creates sessions, closes sessions, verifies VT, and fetches IMEI info from LCS database. By session management, the Communication Agent can read data from VT and write data to VT through unique IMEI, and can manage session interaction through session ID. IMEI will be used as the session ID.

After a connection is accepted, a login message will be sent to Communication Agent, Session Manager will process VT login message, and invoke Application Server LCS Access Bean to dynamically autheaticated IMEI. If IMEI can pass authentication, Session Manager will create an active session for this connection. If the IMEI in login message cannot pass the authentication, Session Manager will reject this connection, no session will be created. And error message will be put into LCS inQueue to notify LCS that an illegal VT whose IMEI is not in the vehicle tracking database tried to access VTS.

After a connection is accepted, if the first message is not the login message, the connection will be dropped directly.

5.2.3 IO Processor

IO Processor is designed to receive data from TCP connection, send data by TCP connection,

> **Glossary**
> fetch *v.* 读取, 提取
> authenticate *v.* 验证
> illegal *adj.* 非法的, 禁用的

encode and decode protocol. It will forward VT message to different inQueue sequences and send messages in outQueue to VT. IO processor is multi-thread based to provide multiple VT connections. The max thread number can be configured in the XML configuration file.

If IO Processor gets a wrong format message, it will reject this record, and put an error message into LCS inQueue to notify LCS an error record is received by VTS. Then the session which received error format record will be closed. Session manager will invoke connection manager interface to close connection.

To support 1,000 connections, IO processor uses the new asynchronous socket API in JDK 1.4. In JDK 1.4, one thread can support several sockets IO. Because the data packet size from/to VT is less than 1,024 bytes, one thread can process at least 50 socket connections. In VTS system, IO processor is designed to dynamically adjust socket connection number in one thread according to the whole TCP connection amount.

5.2.4 JMS InQueue

JMS inQueue can store and manage L0 message from VT, when VT sends L0 message to VTS, IO processor will put L0 message into JMS inQueue, and JMS message processor will process these messages. Because inQueue is used and the L0 message process is isolated from IO processor, the L0 message process time will not block the IO

> **Glossary**
>
> isolate *v.* 隔离, 分离
> block *v.* 拦截

process.

5.2.5 LCS InQueue

LCS inQueue can store and manage messages which should be sent to LCS server, including M0/C0 message from VT and the messages returned from JMS. L0 message in JMS inQueue will be sent to Application Server, but address info returned from Application Server should be put into LCS inQueue. LCS message processor will process all messages in LCS inQueue, send LCS inQueue message to LCS server by TCP connection.

5.2.6 OutQueue

The outQueue can store and manage VT update command message from Application Server or other messages which should be sent to VT.

5.2.7 JMS Message Processor

JMS message processor will manage JMS message, send data request to Application Server by JMS, and receive message returned from Application Server by JMS.

5.2.8 LCS Message Processor

LCS message processor is designed to access LCS. LCS has a listening port. LCS message processor will create a TCP connection to LCS, and at the same time LCS message processor also listens at a specified TCP port to accept connection request from LCS (This means LCS and LCS message processor are Peer to Peer). When TCP connection is created between LCS and LCS message processor,

Glossary

peer *n.* 同位体

Unit 4 Describing Designs

LCS message processor will get message in LCS inQueue, and send to LCS.

There are two LCS connection modes: Long-Connection mode and Multi-Connection mode. Currently, VTS uses the default Multi-Connection mode.

Multi-Connection mode means only VTS gets message from VT, VTS opens the connection with LCS, after VTS sends data to LCS, VTS will close LCS connection. When VTS can not open connection with LCS, it will save the VT data into buffer and send the buffered data when next VT message arrives.

Long-Connection mode means VTS will open connection with LCS at VTS startup, which will close only when VTS shuts down. If the connection is closed by peer or other network problems, VTS will reconnect to LCS. Also, VT data will be buffered when LCS connection is invalid.

5.3 Data Flow

L0: Location Record, L0 data will be sent from VT to VTS, after VTS sends L0 data to GT Server, the GT returned message will be forwarded to LCS by VTS. Then VTS will get return message from LCS, and LCS returned message will be sent to VT by VTS to update VT status.

C0: VT control command, VTS will directly forward C0 data to LCS.

M0: Message command, VTS will directly

> **Glossary**
>
> currently *adv.* 当前
> default *v.* 默认的, 缺省的
> buffer *n.* 缓冲器, 缓冲区

forward M0 data to LCS.

5.4 Protocol

T-TCP protocol will be used between VTS and VT. The protocol between VTS and LCS is T-TCP, too.

5.4.1 VTS/VT Protocol

T-TCP Based.

5.4.1.1 Interactions

The client establishes the TCP connection and logs in. After login, either party may send messages subject to constraints posed by per message acknowledgement and flow pacing delays.

IMEI of VT device will be used as id for authentication. When VT login, it will send its own IMEI to VTS, and VTS will verify this id. Only legal device can login. A 3-bytes app id will be used as application identity. This appid also will be verified by VTS. When VTS sends commands to VT, VT will check the appid to make sure that only legal sources can send commands to VT.

5.4.1.2 Protocol Format

5.4.1.2.1 Login Format

<Aa> <bb> <cc>

0	1	2	3	4	5	6	7	8	9	10	11	12	13	14	15	16	17	18	19	20	21	22

ie: aa<space>350030950880636 with length striped off.

> **Glossary**
>
> strip *v.* 剥离, 去除

The IMEI of GPRS device will be used as deviceid (device identity) for authentication. After login, the device will send the first message right away without waiting for ACK, but after this, everything will go back to normal flow.

5.4.1.2.2 Message Format

<Ll> <id> <space> <body>

0	1	2	3	4	5		N

In VTS, appid is 2 bytes, includes: L0, C0, M0.

L0: Location Record format:

With this appid, reverse geocode is needed.

C0: OTA Command Format:

With this appid, no process is needed. The id will directly pass to LCS.

M0: Message Format:

...

5.4.1.2.3 Acknowledge Format

0	1	2	3		N

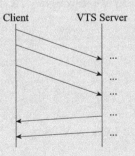

(For the detailed protocol description, please read G-Tech documents.)

5.4.2 VTS/LCS Protocol

The relationship between VTS and LCS is peer to peer; both can be server or client. Login is not needed in current version of LCS. Once TCP connection is created, both sides can send message immediately.

For L0 message, messages from VTS to LCS use T-Protocol like format. Those from LCS to VTS use T-Protocol like format, too. It is the same case with C0 and M0.

There are two LCS connection modes: Long-Connection mode and Multi-Connection mode. Currently, VTS uses the default Multi-Connection mode.

Multi-Connection mode means only VTS gets message from VT, VTS opens the connection with LCS, after VTS sends data to LCS, and VTS will close LCS connection. When VTS can not create connection to LCS, it will save the VT data into buffer and send the buffered data when next VT message arrives.

Long-Connection mode means VTS will open connection with LCS at VTS startup, and only close when VTS shuts down. If the connection is closed by peer or other network problem, VTS will reconnect to LCS. Also, VT data will be buffered when LCS connection is invalid.

If LCS sends command to VT through VTS, LCS will expect ACK back from VT through VTS.

—success

—fail

5.4.2.1 VTS to LCS

Be forwarded from VT to LCS by VTS.

5.4.2.1.1 OTA Command

C0: OTA Command Format

Unit 4 Describing Designs

The original C0 message format from LS to VTS should be: <length> <C0>

<space> <message body>.

Here, when VTS forwards C0 message to LCS, the format <Original message body> should be the same as <message body>.

The OTA Command from VTS to LCS

0	1	2	3	4	5	6	7	22	23	n

5.4.2.1.2 Message Record

M0: Message Format

The original M0 message format from VT to VTS should be: <length> <M0> <space> <message body>.

Here, when VTS forwards L0 message to LCS, the format <Original message body> should be the same as <message body>.

The Message Record from VTS to LCS:

0	1	2	3	4	5	6	7	22	23	n

5.4.2.1.3 Location Record

Location Record Format

The original L0 message format from VT to VTS should be: <length> <L0> <space> <message body>.

Here, when VTS forwards L0 message to LCS, the format <Original message body> should be the same as <message body>. The VTS will get <address> from GT Server. If the position is invalid, or VTS gets error from GT access, VTS should put "No Address Available" as <address>.

The Location Record from VTS to LCS

0	1	2	3	4	5	6	7	22	23		n

> **Glossary**
>
> available *adj.* 可用的, 有效的
>
> append *v.* 追加, 附加

No message length is needed, and the deviceid and address info is added to the original L0 message.

After VTS receives record from VT, VTS will get address message from GT by position, and append address to the end of L0 message, then append "\n" as end mark of message, the message will be sent to LCS. If VTS cannot get address info from GT, VTS will append "No Address Available" as address info.

5.4.2.2 LCS to VTS

Sometimes, LCS will send message to VT via VTS. Only C0 and M0 message will be sent to VT. If LCS sends command to VT through VTS, LCS will expect ACK back from VT through VTS.

—success

—fail

Only C0 and M0 message will be forwarded to VT from LCS.

5.4.2.2.1 OTA Command

C0: OTA Command Format

When VTS gets such message from LCS, VTS will send a C0 message to VT according to

<deviceid>. The message from VTS to VT has the same <message body>, but message length will be re-calculated.

The Message Record from VTS to LCS

0	1	2	3	4	5		m	m+1		n

m = length–14, and the message length forwarded from VTS to VT should be length–15.

5.4.2.2.2 Message Record

M0: OTA Command Format

When VTS gets such message from LCS, VTS will send a M0 message to VT according <DeviceID>. The message from VTS to VT has the same <message body>, but message length will be re-calculated.

The Message Record from VTS to LCS

0	1	2	3	4	5		m	m+1		n

m = length–14, and the message length forwarded from VTS to VT should be length–15.

5.4.3 VTS/GT Protocol

VTS and GT use HTTP and XML as working protocol. A GTHttpRequest class will be used to access GT server. VTS gets longitude and latitude field from L0 message, and sets properties in SessionContext class, and passes SessionContext object as parameter to RevGeoCode class, then gets the address of this location.

6. Application Server

6.1 Overview

Application Server gets message from Communication Agent by JMS,

accesses GT Server, sends L0 message to GT Server and gets returned geocode info from GT Server. After Application Server gets return from GT, it will send it to Communication Agent by JMS so that Communication Agent can send the returned geocode info to LCS. Eventually command from LCS will be sent to VT.

Application Server is based on J2EE, all logical processes (GT Server access or LCS database access) will be handled through EJB. JMS is used for the communication between Application Server and Communication Agent.

6.2 Architecture

Application Server is a group of EJB components, including JMS manager, MDB message processor, GT Access and LCS Access. Communication Agent can send message to Application Server by JMS or directly call EIB to access LCS database.

6.2.1 JMS Manager

...

6.2.2 MDB

...

6.2.3 GT Access Bean

...

6.2.4 LCS Access Bean

...

> **Glossary**
>
> eventually *adv.* 最后，最终

Unit 4　Describing Designs

C Post-reading exercises

Exercise 1

What protocols are to be used between VTS and VT, between VTS and LCS, and between VTS and GT?

Exercise 2

How many modules are included in Communication Agent? What are they?

Exercise 3

Translate the following sentences into Chinese.

a) The existing software will write the records to the existing SQL database and provide the user interface.

b) Session Manager is a working thread which manages all VT–VTS TCP communication sessions.

c) Consumers are empowered with the ability to keep watch of their investments from any computer connected to the Internet.

d) Its main purpose is to manage the incoming message flow to ensure that incoming messages are not lost.

e) The outQueue can store and manage VT update command message from Application Server or other messages which should be sent to VT.

II E-mail & technical writing
Writing

A The length, content and format of the message

E-mails are now considered an important means of communication. In the IT industry, where efficiency and brevity are appreciated, electronic communication has become increasingly popular because of its speed and broadcasting ability. Here is a series of guidelines to help you write effective professional e-mails.

Communicating Professionally and Effectively by E-mail

—The message: length, content and format

√ **Keep the message readable.**

Make your messages brief and to the point. The longer the sentence, the harder it is to follow; complex paragraphs are also more difficult to read. Shorter paragraphs have more impact and are more likely to be read by busy people. Most people can only grasp a limited amount of information within a single paragraph, especially on a computer screen. Accordingly, match the length of the message to the tenor of your communication. For example, if you are only making a simple query, keep your message short and to the point.

√ **Stick to the subject as much as possible.**

Should you need to branch off onto a totally different topic, it is often preferable to send a new message. That allows the recipient the option of filing it separately.

√ **Get your points in order.**

If your e-mail contains multiple messages that are only loosely related, you could number your points to ensure they are all read, inserting an introductory line at the beginning that states how many parts there are in the message. This simple act will reduce the risk that your reader will reply only to the first item that grabs his/her attention. If the points are sufficiently distinct, split them up into separate messages so your recipient can delete, respond, file, or forward each item individually.

√ **Use correct grammar and spelling.**

If your e-mail client has a spell check utility, make use of it. E-mail is all about communication. Poorly-worded and misspelled messages are difficult to read and potentially confusing. Just because e-mail is fast does not warrant careless writing. If your words are important enough to write, they are also important enough to write properly.

√ **Use appropriate and professional language in your communications.**

By using proper e-mail language you will convey a professional image. That will catch your reader's attention, and elicit quicker responses.

√ **Format messages for easy reading.**

White space enhances the look and clarity of an e-mail message, and a blank line only adds one byte to the message. Therefore, always double space between paragraphs. Lengthy messages are almost always read in hard copy form and should be prepared accordingly, e.g., with cover sheets, headers, page numbers, and formatting. These are not suitable for delivery by e-mail.

√ **Use plain text most of the time.**

Be aware that complex formatting may be lost during translation through mail gateways and into mail systems that are not configured to support it.

× **Type your message in all-uppercase.**

A string of words made of uppercase letters is extremely difficult to read. It may also be perceived as yelling, and can be highly annoying and might even

Unit 4 Describing Designs

trigger unwanted flame mail. However, selective use of a short stretch of uppercase may serve to emphasize a point. Therefore, use caution in sending e-mail text in capital letters.

× **Include confidential information in your e-mails.**

Be very careful not to send confidential information such as contracts and credit card numbers in electronic mail. E-mail can be intercepted in transit and a valid credit card number is like cash to an unscrupulous person.

× **Send public "flames"—messages sent in anger.**

Messages sent in the heat of the moment generally only exacerbate the situation and are usually regretted later. Settle down and think about it for a while before drafting a "flame" message.

× **Use HTML or RTF formats as often as possible.**

If your mail program supports fancy formatting (colorful fonts, animated backgrounds and so on) in the mail messages it generates, make certain the recipient has a mail program that can display such messages. Many e-mail clients (and some servers) cannot process such formats.

× **Ask to recall a message.**

After an incorrect message has been sent, you can contact the recipient(s) by phone. It is preferable to sending a follow-up e-mail saying "please don't read the previous message", because this will probably arrive too late. If you cannot phone, send another message to report your mistake, and to apologize. If the message was sent to multiple recipients, a follow-up message is advisable.

B E-mails about design details

1. Gather examples of e-mail from corporate environments and from your personal home files. Compare these two types of e-mail, noting similarities and differences in the length, content, and format. Write an e-mail to your instructor about your findings.

2. Imagine you are writing an e-mail to your colleague who is on the design team in your company, and you have discussed how to improve the FTP client program. You want to tell him/her:

- You agree that the various functions should be realized in more than three components;
- You advise adding FXP functions to the program;
- You ask his/her opinion on the feasibility of developing FXP functions.

Send this e-mail to your partner and your instructor, and review the message received from your partner.

Guide to technical writing

No matter what your current or future job is, writing will be essential to your work because you will have to communicate your technical knowledge to others. Technical documents are a most frequently used type of writing in the IT industry, so training in technical writing will help ensure a smooth career path. Here is a series of guidelines to help you write successful technical documents.

Guidelines to Successful Technical Writing

—Drafting technical descriptions

A technical description is a part-by-part depiction of the components of a program, mechanism, tool, or piece of equipment. Descriptions are a common element in technical writing. Various kinds of technical writing feature descriptions, and several of them consist solely of descriptions, such as requirements specifications, software design specifications, user manuals, and operations manuals.

Follow these guidelines for writing technical descriptions.

Review your pre-writing before you do a first draft.

Before you start drafting, review your brainstorming list. Ask: have you omitted all unnecessary items and included any important information? If not, do so.

Re-organize your outline.

Make sure the items in the outline are organized effectively. To communicate information to your readers, organize the data so readers can follow your train of thought. This is especially important to technical description.

Draft a title precisely stating the topic of your description.

It could be the name of the program, mechanism, tool, or piece of equipment you are describing.

Write a focused statement.

In one sentence, write (1) the name of your topic, (2) its possible functions or your reason for writing the description, and (3) the number of parts comprising your topic. This will eventually become your introduction. For now, however, this sentence can help you organize your draft and maintain your focus.

Compose an introduction.

In the "Introduction" you specify what you are describing, explain its functions or capabilities, and list its primary components.

Draft the text to the best of your ability.

Write quickly without undue concern for grammatical or textual accuracy. Get the information on the paper, focusing on overall organization and a few highlighting techniques. The time to revise and edit comes later.

Write your conclusion according to your purpose.

Your conclusion depends on your purpose in describing the topic. Some options are as follows:

> *Reiteration*   Thus, the xxx program is comprised of the aforementioned components.

Uses   After implementation, you will be able to use this xxx program to...

Comparison   Compared to our largest competitors' programs, the xxx program has more stability than...

The following checklist will help you in writing technical descriptions.

√   Does the technical description have a title noting your topic's name?

√   Does the introduction of the technical description (1) state the topic, (2) mention its functions or the purpose of the mechanism, and (3) list the components?

√   Does the technical description's text use headings to itemize the components in reader-friendly style?

√   Is its detail pictorially precise enough?

√   Are the numbers, calculations and measurements correct?

√   Do you sum up your discussion using appropriate conclusions discussed in this section? Do you use other kinds of conclusions?

D Technical writing exercises

Exercise 1

In two or three paragraphs, discuss the kinds of descriptive writing you expect to do in your field. Be as specific as possible in identifying the subjects, the situations, and the readers.

Exercise 2

After consulting library sources, faculty members, or workers in your field, write a one-page job description of a technical position you hope to hold in five years. Or write a job description of your current job. Include a description of functions, duties, responsibilities, successful experiences and qualifications.

Unit 4 Describing Designs

Exercise 3

Refer to the document in Section 4.3. Describe in your own words the system design of VTS. Pay attention to the guidelines given in this section. After you have finished the composition, check the box next to each guideline if you have incorporated it in your composition.

Unit 5

Detailing Processes
操作细则

- I UML & design specifications
- II E-mail & technical writing

UML & design specifications
Reading

A An overview

Architectural Design

The architectural design shall consist of object and class diagrams in UML notation, describing how each element of the analysis model is to be structured. Classes should be grouped into categories reflecting the logical organization of the system, and into subsystems reflecting the physical organization of the system. Possibilities for concurrency should be identified.

Detailed Design

Software Detailed Design Specification describes the design of the instrument software in sufficient detail to permit code development. A major task of detailed design is to spell out, in detail, the attributes and methods needed by each class described in the architectural design.

B An architectural design specification

Read the following architectural design specification. This is an abridged version, but it sheds much light on what an architectural requirement specification is. For the first time, please only scan the whole document. Keep these questions in mind and try to answer them after scanning. Time limit: 10 minutes.

- What's the primary purpose of this architectural design specification?
- Which design method is used in the design of the software, object-oriented or procedure-oriented?
- Why is the new software necessary?

Architectural Design Specification

Robot Development Project

General Computers Corporation

> **Glossary**
>
> workflow *n.* 工作流, 流程
>
> decomposition *n.* 分解

Abstract

This document describes the architectural design for the Robot Development Project (RDP). This document defines the framework of the solution. This Architectural Design Specification (ADS) is sufficient for the project manager to draw up a detailed implementation plan and to control the overall project during the remaining development phases.

1. Introduction

This section describes the purpose and scope of this document. Also the definitions, acronyms and abbreviations used in this document are described in this section. The last two parts of this section contain references and an overview of the entire document.

1.1 Purpose

The ADS provides a beginning of a solution—by means of a workflow decomposition—for all software requirements as specified in the SRS and defines an object model specifying the components and interfaces between components as they will exist within the Application Programming Interface (API) and the Programming Robots as an Educational and Scientific Tool (PREST) program

which are to be developed during the project. This document is intended for all members of the RDT, as well as senior management, as seen fit by them.

1.2 Scope

The ADS describes the Architectural Design for both the Application Programming Interface (API) and the PREST program. The PREST program must be able to control the movements, and to retrieve and display the positions, of a robotic arm, "NEO Irb-6". The PREST program uses the API, an object-oriented structure which controls the basic movements of the robot as defined in the User Requirements Document. In addition to this, the API's design shall guarantee better extensibility than with the current software, and better control of the current as well as the other robots than the previously described NEO robot.

1.3 Definitions, Acronyms and Abbreviations

ADS	Architectural Design Specification
API	Application Programming Interface
Bit3	Name of the manufacturer of the adapter card. Within this document Bit3 will be used to refer to the adapter card used for this project. The type number: 617
GUI	Graphical User Interface
Irb-6	The type of the NEO robot
iRobot	iRobot Corporation, the client of the RDP
NEO	Brand name of the robot
PREST	Programming Robots as an Educational and Scientific Tool
RDP	Robot Development Project
RDT	Robot Development Team

Glossary

senior *adj.* 高级的
retrieve *v.* 检索，取回
robotic *adj.* 机器人的
guarantee *v.* 担保，保证
extensibility *n.* 可扩展性
brand *n.* 商标，品牌

Unit 5 Detailing Processes

(Continued)

TCP	Tool Center Point. This point is the position on the robot where tools are attached to the robot arm
UML	Unified Modeling Language

1.4 References

G4	"DESIGN PATTERNS: Elements of Reusable Object-Oriented Software" Gamma, Helm, Johnson & Vlissides, Addison-Wesley Longman, 1995
SRS	Software Requirements Specification for RDP, Mary Lee, Jacky Chen, Sam Carter, Robot Development Project, GCC Corporation, 20××
SQAP	Software Quality Assurance Plan, Mary Lee, Jacky Chen, Robot Development Project, GCC Corporation, 20××
SVVP	Software Verification & Validation Plan, Mary Lee, Jacky Chen, Robot Development Project, GCC Corporation, 20××
URD	User Requirements Document for RDP, Mary Lee, Jacky Chen, Sam Carter, Robot Development Project, GCC Corporation , 20××

1.5 Overview

The rest of the ADS contains an overview of the requirements for this project, a decomposition of the software, and an estimate of the feasibility and resources. Section 2 presents an overview of the system that the RDP is intended to produce with 4 UML diagrams and Section 3 places this system in its context (operating environment). Section 4 & 5 provide the decomposition mentioned earlier.

Glossary

reusable *adj.* 可重用的

Section 6 lists all the resources required for the software and includes an estimate of the feasibility of the project.

2. System Overview

The API and the PREST program to be designed and constructed by the RDT are intended to replace the software currently in use by iRobot to control the NEO robot mentioned earlier. The new software is necessary because the old software does not run on new hardware purchased recently by the client, and it is also the intention of the client and the RDP that the new software should be more manageable, more easily extensible and more portable across different robots than the current software. Object orientation must be employed in the design of the software to meet these requirements.

3. System Context

The software will be used primarily by members of the iRobot Corporation for developing a new customer robot model. Other possible users are people with little experience regarding the robot and the API, such as regular end-users. They are considered to be able to use the Windows environment. These users will only use the PREST program.

The program must be able to run on an Intel Pentium III based IBM-compatible PC running Windows 20xx or above, which is the minimum running environment. Microsoft Visual C++ version 6.0 must be used in order to create the API as well

Glossary

intention *n.* 意图，意向
manageable *adj.* 易（可）管理的
portable *adj.* 可移植的
employ *v.* 使用，采用
compatible *adj.* 可兼容的

as the PREST program.

The program is a subsystem of a larger system consisting of the following elements:

- An NEO Irb-6 robot
- A communication medium
- A PC

For further information about the different components please refer to the URD.

The PC and the robot communicate through a Bit3-interface card. The API sends setpoints via the Bit3 interface card to the robot. These setpoints represent the sequence of angles the axles must attain to reach the end position. The robot sends the position of its axles via the Bit3 interface card to the API in the form of a setpoint. Apart from setpoints the API can also send an init command to the robot. This is done by changing the mode of the Bit3 interface card from "user" to "init", then sending the init command and finally setting the mode of the Bit3 interface card back to "user". For more information about the driver of the Bit3 interface card we refer to the Bit3 operations manual.

The PC is equipped with sufficient memory and drive space. Its operating system is Windows 95.

4. System Design
4.1 Design Method

The design method used by the RDT is object-oriented design. In particular we use the Unified

> **Glossary**
>
> sequence *n.* 序列, 顺序
> axle *n.* 轴心
> attain *v.* 达到, 到达

Modeling Language (UML).

4.2 Decomposition Description

This section includes a description of the decomposition of the software into two separate workflows according to the SRS document. This description is given in the form of UML diagrams. The classes in the Class diagram, in section 2, have been rearranged into 2 workflows. This helps to clarify the functions of the system, and the interactions between individual classes. This paves the way for detailed design in which the system will be further decomposed.

According to the URD and SRS, it is viable to separate the workflow into two parts. The first part applies to occasions when a user wants the robotic arm to perform a certain movement directly through the GUI, and the second part applies to occasions when a user wants the robot to perform as prescribed in a movement file. This decomposition has been chosen to bundle classes with a certain goal together and keep the interfaces as few as possible.

4.3 Movement Control

The use case and sequence diagrams below show the events if the user wants to let the robot perform a movement.

In the diagrams above, it is assumed that the robot has already been initialized. To make the robot move, the user selects the parameters via the GUI, PREST handles this event and employs the Add*Move (*: wildcard) method from the current

Glossary

pave v. 铺
occasion n. 场合
prescribe v. 指定, 规定
bundle v. 包, 捆
event n. 事件
wildcard n. 通配符
instance n. 实例

instance of the Toolkit class. This method creates the proper instance of the Movement class and adds this movement to a list kept in an instance of the Movement Sequence class.

Glossary
proper *adj.* 正确的，合适的
prompt *v.* 提示

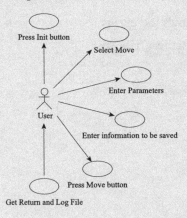

The PREST program then prompts the user that a movement is about to be executed and asks for confirmation. This is indicated by the second arrow from User to GUI. PREST handles this event and calls the run method in the current instance of the Toolkit class. This run method then calls the calc method of the instance of Movement Sequence that keeps the selected movement. This calc method calls the calculate methods for all the movements in the sequence. The calculate method of this Movement instance calculates a set of speed profiles and the setpoints to be sent to the robot.

After the speedprofiles and setpoints have been calculated, the run method is called for all Movement instances in the sequence. This method then sends the calculated setpoints to the current instance of the Robot class. Finally, after all setpoints have been sent to the Robot instance (i.e. when the movement has ended) a notification is sent to the observer, so that the PREST instance knows the movement has ended.

4.4 Movement File Control

The sequence diagram below shows the events when a user wants the robot to execute a file of movements.

Glossary

interpreter *n.* 解释程序
trigger *v.* 启动, 触发
option *n.* 选项

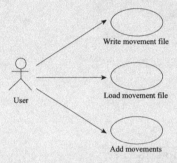

To make the robot perform a sequence of movements the user selects a file in the PREST program. This results in a call of the load Movement File method in the current instance of PREST. After the file has been read, a new instance of Interpreter is created and the file is passed to this instance. This interpreter generates events causing the current PREST instance to call the appropriate Add*Move method of the current Toolkit instance.

Further adding of movements, calculating of setpoints and speedprofiles and sending of these setpoints to the robot work in the same way as is explained in the text below the first sequence diagram. The actual sending of setpoints to the robot is triggered by a menu option "Execute movement file" in the movement menu of the PREST program.

C A sample of detailed design specification

Read the following detailed design specification. This is an abridged version, but it sheds much light on what a detailed requirement specification is. For the first time, please only scan the whole document. Keep these questions in mind and try to answer them after scanning. Time limit: 10 minutes.

- What's the main difference between architectural design and detailed design?
- What does this document mainly describe?
- What tools are used for the detailed design according to this document?

> Glossary
>
> concurrently *adv.* 并发地

Detailed Design Specification

Robot Development Project

General Computers Corporation

Abstract

This document describes the detailed implementation for the Robot Development Project (RDP). It contains two parts. The first part defines design and coding standards and tools. The second part contains detailed implementation information for each component. The second part of this document is produced concurrently with the detailed design, coding and testing.

Table of Contents

    Abstract

    Table of Contents

    Document Status Sheet

    Document Change Record

Part 1 General Description

1. Introduction

   1.1 Purpose

   1.2 Scope

   1.3 Definitions, Acronyms and Abbreviations

   1.4 References

   1.5 Overview

2. Project Standards, Conventions and Procedures

   2.1 Design Standards

   2.2 Documentation Standards

   2.3 Naming Conventions

   2.4 Programming Standards

   2.5 Software Development Tools

   2.6 Procedure of the API Classes

Part 2 Component Design Specifications

3. Component 1

4. Component 2

5. Component 3

6. Component 4

   For every component:

   Type

   Purpose

   Function

   Subordinates

   Dependencies

Glossary

subordinate *n.* 从属物件

Interfaces

Resources

References

Processing

Data

Document Status Sheet		
Document Title	Detailed Design Specification	
Document Reference Number	RDP-DDS/0.5.0	
Issue	**Revision**	**Date**	**Reason for change**
	0.1.0	April 28 20xx	Creation of document
	0.2.0	June 8 20xx	Processing review items
	0.3.0	June 16 20xx	Second review of the document
	0.4.0	June 30 20xx	Created after first external review
	0.5.0	July 6 20xx	Remarks made on the second external review on July 5

Document Change Record			

Part 1 General Description

1. Introduction

This section describes the purpose and the scope of this document. Also the definitions, acronyms and abbreviations used in this document are described in this section. The last two parts of this section contain references and an overview of the entire document.

1.1 Purpose

This document defines design, coding standards and tools. The design from the AD phase is detailed in the DD phase, so that it can be implemented directly. Programmers should obey the standards and use the tools mentioned in Part 1 of this document. During the implementation of the design, documentation for each component is produced. The results are contained in Part 2. All members of the RDP team as well as the senior management and the more experienced user should read this document.

1.2 Scope

The DDS describes the Detailed Design for both the Application Programming Interface (API) and the PREST program. The PREST program must be able to control and simulate the movements, and retrieve and display the positions, of a robotic arm, "NEO Irb-6". The PREST program uses the API, an object-oriented structure which controls the basic movements of the robot as defined in the URD. In addition to this, the API's design will guarantee better extensibility than the current software, both

Glossary

obey v. 服从, 遵守
simulate v. 模拟, 仿真

with respect to the current robot as to the control of other robots than the NEO robot.

1.3 Definitions, Acronyms and Abbreviations

ADS	Architectural Design Specification
API	Application Programming Interface
Bit3	Name of the manufacturer of the adapter card. Within this document Bit3 will be used to refer to the adapter card used for this project. The type number: 617
DDS	Detailed Design Specification
GUI	Graphical User Interface
Irb-6	The type of the robot NEO
iRobot	iRobot Corporation, the client of the RDP
NEO	Brand name of the robot
PREST	Programming Robots as an Educational and Scientific Tool
RDP	Robot Development Project
RDT	Robot Development Team
TCP	Tool Center Point. This point is the position on the robot where tools are attached to the robot arm.
UML	Unified Modeling Language

1.4 References

ADS	Software Quality Assurance Plan, Mary Lee, Jacky Chen, Robot Development Project, GCC Corporation, 20xx
G4	"DESIGN PATTERNS: Elements of Reusable Object-Oriented Software" Gamma, Helm, Johnson & Vlissides, Addison-Wesley Longman, 1995
SRS	Software Requirements Specification for RDP, Mary Lee, Jacky Chen, Sam Carter, Robot Development Project, GCC Corporation, 20xx
SQAP	Software Verification & Validation Plan, Mary Lee, Jacky Chen, Robot Development Project, GCC Corporation, 20xx
SVVP	Software Verification & Validation Plan, Mary Lee, Jacky Chen, Robot Development Project, GCC Corporation, 20xx
URD	User Requirements Document for RDP, Mary Lee, Jacky Chen, Sam Carter, Robot Development Project, GCC Corporation, 20xx

1.5 Overview

The DDS consists of two parts. The first is about project standards, conventions and procedures that should be used by the implementers. Part 2 describes for each component its design specifications. There are two appendices in this document. Appendix A lists all source code that is produced during this phase. The Software Requirements vs. Components Traceability matrix is contained in appendix B.

2. Project Standards, Conventions and Procedures

2.1 Design Standards

The design method used is object-oriented in design. The Object Modeling Technique (OMT) is used in particular for our project. The method of stepwise refinement and structured programming are used as described by E.W. Dijkstra. This means:

- Use only selection, sequence and iteration statements.
- Do not use jumps (goto).
- Use hierarchical decomposition: if nesting becomes too deep, define a new lower-level module.

2.2 Documentation Standards

The tools used for the detailed design are:

- IBM Rational Rose 20xx
- Macromedia Dreamweaver MX
- Microsoft Visual C++

Glossary

stepwise *adj.* 步进式, 逐步的

refinement *n.* 精化, 精细化

The standard module header can be found in section 2.4.

2.3 Naming Conventions

Variables will be documented in the following style:

int max Speed; // the maximum speed of the upper arm

Exceptions are in capital letters, ending with "EXCEPTION". The words in an exception separated by underscores, for instance: OUT_OF_BOUNDS_ EXCEPTION.

Class names will be documented in the same way as variable names, except that the first letter must be a capital letter, for instance: Robot.

File Type	Extension	Commentary
documentation	.html	
source code	.cpp	Should be the same name as the class name.
header files	.h	Should be the same name as the class name.

2.4 Programming Standards

During the RDP the following standards on behalf of coding and commentary constructs and layouts are used. This means that the code written by RDT members must comply with these standards.

- standard file headings

(omitted)

- standard class definitions

> **Glossary**
> underscore *n.* 下划线, 底线
> commentary *n.* 注释
> construct *n.* 构造, 构成
> layout *n.* 布局, 格式
> comply *v.* 遵守, 依从

class -class name- (: (public, private) -parentclass-)

{

private:

-tab- -private declarations-

public:

-tab- -public declarations-

};

- standard method declarations

-return type- -method name- (-parameters-)

// pre-condition: -pre-condition description-

// post-condition: -post-condition description-

// returns: -return description-

- standard method definitions

-return type- -classname- :: -method name- (-parameters-)

// pre-condition: -pre-condition description-

// post-condition: -post-condition description-

// returns: -return description-

{

- implementation -

}

- standard variable names

Variable names in US English. If a variable name is a combination of several words, these words are separated by capital letters. For instance: maxInt.

Glossary

declaration *n.* 声明，说明

- commentary language

All commentary is written in US English.

2.5 Software Development Tools

We use IBM Rational Rose 20xx for the object-oriented design, in particular for the drawing of the OMT diagrams and for the sequence diagrams. For the development of the software Microsoft Visual C++ is used. As the client is also familiar with this tool, this tool is our major resource editor. It consists of a source code editor, compiler, debugger, and linker.

For our documentation HTML is used. Documentation is produced with several editors and Macromedia Dreamweaver MX which is an HTML editor.

2.6 Procedure of the API Classes

The diagram of the API and its classes explores the state transitions of the API through various types of movements. This diagram is useful for further component decomposition, which will be done in the second part.

The API gets a movement instruction, calculates the speed profile, verifies it and then calculates the setpoints of this move and verifies them. Next if the robot moves, the setpoints are sent to the Bit3. If all setpoints are done, the information about the move is saved into the log file. The PREST program calls during the move for setpoints and for a top and side view. The robot can always be halted by the interrupt

> **Glossary**
>
> halt v. 暂停, 异常终止

button.

Part 2 Component Design Specifications

3. Component 1

3.1 Type

Hardware abstraction layer

3.2 Purpose

SR01, SR11, SR15, SR20–26, SR28, SR30

3.3 Function

All robot-specific information of the API is provided to the rest of the program from this component. Any external resources to the software must report itself to the Robot class as an observer in order to receive information.

3.4 Subordinates

- The Robot class, which controls the information flow between the components and between the components and the external resources.

- The Subject class, which exists to regulate where information about the robot is sent.

- The Setpoint and Position classes, which are container classes with which setpoints in robot coordinates and positions in Cartesian coordinates may be communicated across the API.

3.5 Dependencies

Other than initialization, no operations are

Glossary

interrupt *n.* 中断
regulate *v.* 管理, 控制
initialization *n.* 初始化
entity *n.* 实体

required before this component may be used. However, it is necessary for entities which want to receive data from the component to register themselves as observers of the component and its classes.

3.6 Interfaces

The Setpoint and Position classes are used as communication classes of the Robot class. The Subject class provides the Robot class with a list of observers which must be notified if the robot changes.

Instances of the Robot class serve as abstractions of physical robots which may be controlled by the software. In this project there is only one child of the Robot class, the NEO Irb-6 class. This class initializes the simulation part in such a way that the classes resemble the NEO Irb-6 robot and that all bounds (upperbound, lowerbound, max speed, max acceleration) of the joints are correct.

3.7 Resources

This component requires the observers of Component 3 for the actual sending of setpoints to the robot or to a file.

3.8 References

-

3.9 Processing

All processing in this component is distributed to other Components. This Component gets a request and gets the answer of that request from another Component. The processing of Setpoints is done in Component 3 and information about the

> **Glossary**
>
> abstraction *n.* 抽象
> distribute *v.* 分布，配送

robot is stored in Component 2.

3.10 Data

The internal data of the component are a list of observers of the Robot, all the relevant memory addresses required for access to the driver, and the Bit3 card. Further an instantiation of the Setpoint class is used to keep track of the actual Setpoint. The necessary initialization for the NEO Irb-6 robot used by the client is done in the NEO Irb-6 class.

4. Component 2

4.1 Type

Abstraction layer for the various parts of the robot.

4.2 Purpose

SR11, SR15, SR20–25

4.3 Function

This component contains an abstraction of the parts of the robot. It is responsible for the conversion of setpoints to positions and for drawing of the robot on a given Device Context.

4.4 Subordinates

The Part class with its children (TCP, Robot Arm and Joint).

4.5 Dependencies

The instances that will consist of the several children of the Part will be created by the Robot class. Therefore an instance of Robot must exist

Glossary

instantiation *n.* 例化, 示例
conversion *n.* 转换, 换算
virtual *adj.* 虚拟的

before any instance of a Part can exist and thus before any method from Component 2 can be called.

4.6 Interfaces

The Part class contains a number of virtual methods which are implemented in the children of Part (Robot Arm, Joint and TCP). We will discuss these methods first.

4.7 Resources

The only resources needed by this component (aside from the instance of Robot which creates a chain using this component) are sufficient memory to store the chain representing some robot and a valid CDC for every invocation of the draw(CDCs, t) method sequence.

4.8 References

Not applicable.

5. Component 3

5.1 Type

Observers of the robot.

5.2 Purpose

SR03-04, SR11, SR14-15, SR18-25, SR28-29

5.3 Function

The observer class and its children react if the state of the subject (Robot class) changes. Furthermore the driver class is also used by the Robot class to get the position of the robot and to initialize and halt the robot.

5.4 Subordinates

- The Observer class, which is an abstract class with an abstract update method. This method is called if the subject—Observer—changes its state.
- The Driver class, this class communicates with the Bit3 interface card. It's an observer that sends the new setpoint to the Bit3 interface

card if the setpoint changes in the subject. It also retrieves the position of the robot and it can initialize and stop the robot.

- The LogFile class, which writes setpoint to a file. One can choose if only the setpoints from the Robot class or only the Setpoint of the real robot or both are written to file. One can also choose if all setpoint must be written or if only the values for certain axles must be written.

5.5 Dependencies

These classes need to register with the subject (Robot class) if they want to be informed whether the state of the subject changes.

5.6 Interfaces…

(omitted by editor)

6. Component 4

6.1 Type

Functional core of the API.

6.2 Purpose

SR01-02, SR04-13, SRD16-17, SR23-30

6.3 Function

This component is where all the calculations for movements are made. Furthermore the calculated setpoints are sent to the robot.

6.4 Subordinates

ADDC3.Movement, ADDC3.CMovementNode, ADDC3.MovementSequence, ADDC3.Movement, ADDC3.Toolkit.

6.5 Dependencies

None.

6.6 Interfaces

The execution of this component is started when a program built on top of the API calls one of the functions of the toolkit-class. In this way, movements are programmed, started and halted. Also this component interfaces

Unit 5 Detailing Processes

with Position- and Setpoint-classes (Component 1) in order to perform transformations of the one to the other, and store calculated movements. This component is also attached to the Robot class, to get information about the robot, needed for calculations and to send commands to it.

6.7 Resources

Memory.

6.8 References

Not applicable.

6.9 Processing

...

(omitted by editor)

D Post-reading exercises

Exercise 1

How many types of UML diagrams are mentioned in the two specifications? What are they?

Exercise 2

What standards should the members conform to when they are programming?

Exercise 3

Fill in the table according to component design specifications.

Component	Type	Function
Component 1		
Component 2		
Component 3		
Component 4		

Exercise 4

What are the main dependencies of every component? Please try to depict them in Chinese briefly.

Exercise 5

Translate the following sentences into Chinese.

a) This Architectural Design Specification (ADS) is sufficient for the project manager to draw up a detailed implementation plan and to control the overall project during the remaining development phases.

b) Object orientation must be employed in the design of the software to meet these requirements.

c) The program must be able to run on an Intel Pentium III based IBM-compatible PC running Windows 2000 or above, which is the minimum running environment.

d) The software will be used primarily by members of the iRobot Corporation for developing a new customer robot model.

II E-mail & technical writing
Writing

A Manner & tone

E-mails are now considered an important means of communication. In the IT industry, where efficiency and brevity are appreciated, electronic communication has been increasingly popular because of its speed and broadcasting ability. Here is a series of guidelines to help you write effective professional e-mails.

Communicating Professionally and Effectively by E-mail

—The message: manner & tone

√ **Mind your manners and use courteous expressions frequently.**

What three words have a total of only 14 letters yet carry a great deal of meaning? People may not notice these words, but if you forget to use them, you will give the impression of being disrespectful and ungrateful. These very powerful words are "Please," and "Thank You." Polish your manners with these words.

√ **Use a positive, friendly and appropriate tone.**

Be aware of the tone of your communication. You should be perceived as being respectful, friendly, and approachable. You should not sound curt or demanding. When you communicate with another person face to face, 93% of the message is non-verbal. E-mail, however, has no body language. The reader cannot see your face or hear your tone of voice, so choose your words carefully. Put yourself in the recipient's place and try to imagine how your words may come across in Cyberspace. That message still represents you to your recipient, and it should represent you well. At the same time, be careful not to make statements which could be interpreted as official positions of your organization.

√   **Maintain professionalism.**

Whether you are sending a message to a colleague or are transmitting a proposal to a potential client, certain e-mail formalities are required. Just because an e-mail message is more immediate than other forms of communication does not suggest that the tone can be more casual. Your e-mail message should carry the same degree of professionalism as a document that is distributed on your organization's letterhead.

√   **Tailor your message to the receiver.**

Although your message is a reflection of you, the style of your message should be tailored to the person receiving it. For instance, if you have a close rapport with the person and your message deals with a "light" topic, a more casual tone is acceptable. However, if your relationship is more formal and your topic is more serious in nature, a corresponding tone should be used. When unsure of the appropriate tone, write the way you would talk to the person.

×   **Write or send e-mails that will backfire.**

E-mail is not meant for certain kinds of humor, sarcasm, ironic asides or critical assessments. This medium is far too literal and fast moving for nuances. Do not do it.

Certain topics should be entirely forbidden. In business, never use e-mail to send off-color jokes or to comment on anyone's sexual, racial, religious or ethnic characteristics or practices. Never send messages about someone's age or disability. Never trash a colleague's professional capabilities or job performance or history. Some of this could be legally actionable. Any of it could severely damage your reputation and business.

×   **Write what you wouldn't say.**

We've all received our fair share of e-mail that is annoying, offensive, weird, unnecessary, confusing, or incomprehensible. Use these experiences as bad examples. Put yourself in a recipient's place and, before hitting "send", ask yourself: Would I want to receive this message? This method works.

B E-mails describing UML diagrams

1. Gather two sample e-mails from the corporate world and from your home files. Compare these two types of e-mails. What similarities or differences in tone are evident in both types of correspondence? If the tones in both corporate and home-use e-mails are similar, is that appropriate? Could this lead to problems? Write an e-mail to your instructor to explain your answers.

2. Imagine you are writing an e-mail to a colleague. The recipient is a new member of the development group, and does not know much about UML diagrams. You want to send him/her an e-mail with:

- A brief introduction to the Unified Modeling Language;
- A description of a sample Use Case Diagram in the document in Section 5.3;
- Ask him/her if there are any questions.

Send this e-mail to your partner and your instructor, and review the message sent from your partner.

C Guide to technical writing

No matter what your current or future job is, writing will be essential to your work because you will have to communicate with others about your technical knowledge. Technical documents are a most frequently used type of writing in the IT industry, so training in technical writing will help ensure a smooth career path. Here is a series of guidelines to help you write successful technical documents.

Guidelines to Successful Technical Writing

—Writing short reports

We have learned about drafting general technical documents in the previous unit. The next three units will focus on features of specific technical documents.

In the professional world, decision makers rely on reports to present ideas and facts. Every long report and numerous short reports lead to informed decision making.

Next to correspondence, short reports are the most frequently written documents in the job. The most common types of reports include feasibility reports, progress reports, activity reports, study reports, and performance appraisal reports. When writing short reports, follow these general guidelines:

Use good organization.

Every report should contain four basic parts: heading, introduction, discussion, and conclusion/recommendations.

Develop your report properly.

First, keep your audience and objectives in mind. Second, provide sufficient information such as who participated when the event occurred, where the event took place, why you are writing this report, and what conclusions or recommendations you have reached. Finally, quantify your points clearly and precisely with graphics where appropriate.

Adopt a correct language style.

Style encompasses conciseness, simplicity, and highlighting techniques. You achieve conciseness by eliminating wordy phrases, simplicity by avoiding dated words and phrases, and highlights by use of headings, italics, figures and tables.

The purpose of short reports is to communicate objectively and precisely in memo form, letter form, in a prepared form, or in a miscellaneous form.

Letter Reports

Letter reports provide information often to people outside the organization.

They are typically informational or recommendation reports.

Follow the standard letter format.

Insert additional headings where useful.

Two format additions to a letter report are: (1) a subject heading, placed right below the salutation, and (2) other sub-headings designed to segment your letter into specific subsets of information.

Be sure your letter report retains the personal "you" perspective.

Memo Reports

Memos are the major form of written communication within an organization. This popular method of communication can touch on any topic important to its operation.

Write a heading with the organization name, sender, recipient, subject, and date.

If the memo report is longer than one page, then list the recipient, date and page number on following pages.

Most memos simply end after the final point.

Memos don't require a complimentary close or signature.

Progress Reports

Large organizations depend on progress (or status) reports to keep track of activities, problems, and progress of various projects. To give management the answers it needs, progress reports must, at a minimum, answer these questions:

1. How much has been accomplished since the last report?

2. Is the project on schedule?

3. If not, what went wrong?

    A) How was the problem corrected?

    B) How long will it take to get back on schedule?

4. What else needs to be done?

5. What is the next step?

6. Are there any unexpected developments (other than schedule problems)?

7. When do you anticipate completion? Or, on a long project, when do you anticipate completion of the next phase?

Make sure your progress report answers the above questions.

Include the topic about which you are reporting and the reporting interval in the question.

Bring your readers up to date with background data or a reference to previous reports.

Choose an optimum report structure and format. Many organizations have forms for organizing progress reports, so there isn't best format.

The format and organization must fit your purpose, audience, and situation.

Prepared-form Reports

To streamline communications and keep track of data, many companies use prepared forms for short reports.

Follow the guidelines provided by prepared form.

If you complete the form correctly, you are sure to satisfy the readers' needs.

Make your language simple and clear enough for quick review.

One major purpose of a prepared form is to standardize data reported from various sources. Tailor your report for rapid processing.

If necessary, attach your own statement explaining certain items in the form.

Some prepared forms are limited in space, so you may need to give additional explanations.

Unit 5   Detailing Processes

D Technical writing exercises

Exercise 1

You would like to see some changes made in this course to better suit your career plans. Perhaps you feel too much emphasis is placed on e-mail writing, too little on reports. Or maybe there is too much lecturing and not enough discussion. Write a letter report to your instructor, justifying the reasons for the changes you propose. Remember, you must illustrate specific benefits for you and your classmates.

Exercise 2

Think of an idea you would like to see implemented on your campus or on your job. Write a short feasibility report, persuading your audience that your idea is worthwhile.

Unit 6

Documenting Your Work

文档制作

I Source code documentation conventions

II E-mail & technical writing

Source code documentation conventions
Reading

A An overview

Code Documentation (Software Documentation)

Code documentation is what most programmers mean when using the term software documentation. When creating software, code alone is insufficient. There must be some text along with it to describe various aspects of its intended operation. This documentation is usually embedded within the source code itself so it is readily accessible to anyone who may be traversing it.

This writing can be highly technical and is mainly used to define and explain the API's, data structures and algorithms. For example, one might use this documentation to explain that the variable m_name refers to the first and last names of a person. It is important for the code documents to be thorough, but not so verbose that it becomes difficult to maintain them.

Often, tools such as Doxygen, javadoc, ROBODoc, POD or TwinText can be used to auto-generate the code documents—that is—they extract the comments from the source code and create reference manuals in such forms as text or HTML files. Code documents are often organized into a reference guide style, allowing a programmer to quickly look up an arbitrary function or class.

Many programmers really like the idea of auto-generating documentation for various reasons. For example, because it is extracted from the source code itself (for example, through comments), the programmer can write it while referring to his code, and can use the same tools with which he created the source code to make the documentation. This makes it much easier to keep documentation up-to-date.

Unit 6 Documenting Your Work

Of course, a downside is that only programmers can edit this kind of documentation, and it depends on them to refresh the output (for example, by running a cron job to update the documents nightly). Some would characterize this as a pro rather than a con.

B A naming and code documentation guide

Read the following guide about source code documentation. This document sheds much light on how a programmer is expected to document his/her code in General Computers Corporation. For the first time, please only scan the whole document. Keep these questions in mind and try to answer them after scanning. Time limit: 10 minutes.

- Where shall we go if we want to know about implementation comments?
- When should we use block comments in code documentation?
- How many types of documentation comments are included in this

Naming and Code Documentation Guide

General Computers Corporation

Table of Contents

1. Naming Conventions
   1.1 Class Files and Header Files
   1.2 Function Names
   1.3 Class Names
   1.4 Variable Names
2. Code Documentation
   2.1 Implementation Comment Formats
      2.1.1 Block Comments
      2.1.2 Single-line Comments

Glossary

convention *n.* 约定, 规范

function *n.* 函数

2.1.3 **Trailing Comments**

2.1.4 **Out Comments**

2.2 **Documentation Comments**

2.2.1 **Copyright Information**

2.2.2 **Module Comments**

2.2.3 **Revision History**

2.2.4 **Code Specifications**

2.2.5 **Sources**

Glossary

trailing *adj.* 拖尾的

rule of thumb 经验法则

potential *adj.* 可能的, 潜在的

prefix *n.* 前缀

1. **Naming Conventions**

1.1 **Class Files and Header Files**

The name of files can be more than eight characters, with a mix of upper cases and lower cases. The name of files should reflect the content of the file as clearly as possible.

As a rule of thumb, files containing class definitions and implementations should contain only one class. The name of the file should be the same as the name of the class. Files can contain more than one class when inner classes or private classes are used.

Special care should be given to the naming of header files because of potential conflicts between modules. If necessary, a module prefix could be added to the filename. For example, if two modules have a garbage collector class: Database and Window, the files could be named: "DBGarbage Collector.H" and "WINDOWGarbageCollector.H".

The following table illustrates the standard file extensions in use.

Extension	Description
.C	C source files
.CPP	C++ source files
.H	C/C++ header files
.INL	C++ inline function files

> **Glossary**
>
> concatenate v. 连接, 并置
> routine n. 例程, 例行程序
> procedure n. 过程, 程序
> cohesion n. 内聚度
> imply v. 暗示, 意味

1.2 Function Names

Class member functions follow Java conventional naming. The function name is a concatenated string of words with the first letter of all words capitalized except for the first one. For example: isMemberSet, printReport, consume.

Functions exported from DLLs, that are not in a class or a namespace should include an uppercase abbreviation of the module name. For example: DEBUGTrace or DBGTrace.

A good name for a routine clearly describes everything the routine does. Here are guidelines for creating effective routine names.

For a procedure name, use a strong verb followed by an object. A procedure with functional cohesion usually performs an operation on an object. The name should reflect what the procedure does, and an operation on an object implies a verb-plus-object name. printReport(), calcMonthlyRevenues(), and repaginateDocument() are samples of good procedure names.

In object-oriented languages, you don't need to include the name of the object in the procedure name because the object itself is included in the call.

For a function name, use a description of the return value. A function returns a value, and the function should be named for the value it returns. For example, cos(), nextCustomerID(), printerReady(), and currentPenColor() are all good function names that indicate precisely what the functions return.

Avoid meaningless or wishy-washy verbs. Some verbs are elastic, stretched to cover just about any meaning. Routine names like handleCalculation(), performServices(), processInput(), and dealWithOutput() don't tell you what the routines do. At the most, these names tell you that the routines have something to do with calculations, services, input, and output. The exception would be when the verb "handle" is used in the specific technical sense of handling an event.

The properties of a class should be accessible through *getter* and/or *setter* methods. Those methods should always be declared like this:

public boolean get();

public void set(boolean a);

For boolean properties, the use of "is" can replace the "get":

Glossary

precisely *adv.* 精确地, 明确地

wishy-washy *adj.* 乏味的

elastic *adj.* 灵活的

stretch *v.* 延伸, 扩展

```
public boolean is();
```
Examples:
```
Color getCurrentPenColor();

void setCurrentPenColor(Color c);

boolean isPrinterReady();

void setPrinterReady(boolean ready);
```

1.3 Class Names

Classes follow Java conventional naming. The name is a concatenated string of words with the first letter of all words capitalized. For example: FocusEvent, DeviceContext, Customer.

Interface names follow COM conventional naming. The name is a concatenated string of words with the first letter of all words capitalized. A capital "I" is used as a prefix. For example: IWindowModel, IUnknown.

1.4 Variable Names

Variable names are a concatenated string of words with the first letter of all words capitalized except for the first one. The name chosen should clearly represent the content of the variable. For example: windowHandle, eventConsumed, index.

All variables that are a member of a class should have an "m_" prefix. For example: m_windowHandle, m_eventConsumed, m_index.

Constant (static final for Java) variables make exception to all these rules. They should follow the same standard as #define statements in C++. Constant variables should be named using capitalized names separated by an underscore character ('_'). For example: MAX_ARRAY_LENGTH, SIZE_PROPERTY_NAME.

2. Code Documentation

When applicable, all source documentation should be in a format compatible with the generic formatting defined by the original programming

language. However, it is important that the formatting codes included in the comment blocks do not overwhelm the comment block. Don't forget that the comment block is meant to be read in the source first.

Glossary

generic *adj.* 一般的, 通用的
overwhelm *v.* 淹没, 压倒
delimit *v.* 定界限, 限定
readily *adv.* 容易地

Programs in GCC Corporation have two kinds of comments: implementation comments and documentation comments. Implementation comments are delimited by /*...*/, // (as in C/C++ environment) or ' (as in VB environment). Documentation comments should be delimited by /**...*/ (as prescribed in JavaDoc formatting) or ' (VB).

Implementation comments are means for commenting out code or for comments about the particular implementation. Documentation comments are meant to describe the general information of the code and the specification of the code, from an implementation-free perspective to be read by developers who might not necessarily have the source code at hand.

Comments should be used to give overviews of code and provide additional information that is not readily available in the code itself. Comments should contain only information that is relevant to reading and understanding the program. For example, information about how the corresponding package or procedure is built or in what directory it resides should not be included as a comment.

Discussion of non-trivial or non-obvious design decisions is appropriate, but avoid duplicating information that is present in the code. It is too easy for redundant comments to get out of date. In general, avoid any comments that are likely to get out of date as the code evolves.

Comments should not be enclosed in large boxes drawn with asterisks or other characters. Comments should never include special characters such as form-feed and backspace.

Inline comments should be made with the "//" comment style and should be indented at the same level as the code they describe. End-of-line comments should be avoided with the exception of function parameters.

2.1 Implementation Comment Formats

Programs can have four styles of implementation comments: block, single-line, trailing and out comments.

2.1.1 Block Comments

Block comments are used to provide descriptions of files, methods, data structures and algorithms. Block comments may be used at the beginning of each file and before each method. They can also be used in other places, such as within methods. Block comments inside a function or method should be indented to the same level as the code they describe.

A block comment should be preceded by a

Glossary

non-trivial *adj.* 重要的
duplicate *v.* 复制，重复
redundant *adj.* 冗余的
evolve *v.* 发展
asterisk *n.* 星号，星标
indent *v.* 缩进

blank line to set it apart from the rest of the code. For example:

```
/*
================================
void BCMenu::DrawItem(LPDRAWITEMSTRUCT)
----------------------------------------
Called by the framework when a particular item needs to be drawn. We overide this to draw the menu item in a custom-fashion, including icons and the 3D rectangle bar.
================================
*/
```

Glossary

apart *adv.* 分开, 离开
differ *v.* 相异, 不同

The example above exists in C/C++ environment. In Java and VB, the comment formatting differs.

```
/*
* Returns the resource of a given name associated with a
* particular class (never null), or null if none can be
* found.
*/
'NOTE: The following procedure is required by the Windows Form Designer.
'It can be modified using the Windows Form Designer.'Do not modify it using the code editor.
```

2.1.2 Single-line Comments

Short comments can appear on a single line indented to the level of the code that follows. If a comment can't be written in a single line, it should follow the block comment format (see section 2.1.1). A single-line comment should be preceded by a blank line. Here are two examples of a single-line comment in Java and VC codes:

```
if (condition) {
/* Handle the condition. */
...
}
if (...) {
// this is a text control: set up font and colors
if (...) {
// first time init: create font
...
}
}
```

2.1.3 Trailing Comments

Very short comments can appear on the same line as the code they describe, but should be shifted far enough to separate them from the statements. If more than one short comment appears in a chunk of code, they should all be indented to the same tab setting. Here's an example of a trailing comment in Java and C# codes:

Glossary

shift *v.* 移位

chunk *n.* 块, 程序块

```
if (a == 2) {
return TRUE;     /* special case */
} else {
return isPrime(a);     /* works only for odd a */
}
try     {// This operation is skipped in partial trust scenarios.
using (...) {
...
}
}
```

> **Glossary**
>
> delimiter *n.* 定界符, 分隔符
>
> consecutive *adj.* 连续的

2.1.4 Out Comments

The // comment delimiter can comment out a complete line or only a partial line. It can also be used in consecutive multiple lines for commenting out sections of code. An example in Java code follows:

```
//if (bar > 1) {
//
// // Do a triple-flip.
// ...
//}
//else{
// return false;
//}
```

Unit 6 Documenting Your Work

2.2 Documentation Comments

Programs can have five types of documentation comments: copyright information, module comments, revision history, code specification, and sources. The information provided by documentation comments delivers very important messages to other developers and users. Adequate documentation comments help a lot in building user-oriented documents in the later stages of projects.

2.2.1 Copyright Information

This section provides the copyright and licensing information of the following code. For example:

```
/*
 * @(#)RobotArm.java 1.82 99/03/18
 *
 * Copyright (c) 1995-2005 GCC Corporation (China).
 * 999 Tsinghua Science Park, Haidian Dist., Beijing,
 * 100084, P.R.C.
 * All Rights Reserved.
 *
 * This software is the confidential and proprietary
 * information of the GCC Corporation (China).
 * You shall not disclose such Confidential information and
```

> **Glossary**
>
> adequate *adj.* 适当的, 足够的

```
    * shall use it only in accordance with
the terms of the
    * license agreement you entered into
with GCC.
    */
```

2.2.2 Module Comments

Module headers are block headers placed at the top of every implementation file. The block comment should contain enough information to tell the programmer if he/she reached his/her destination.

Glossary

destination *n.* 目的，目标

```
    /*
    ** FILE: filename.cpp
    **
    ** ABSTRACT:
    ** A general description of the
       module's role in the overall software
       architecture, what services it provides
       and how it interacts with other
       components.
    **
    ** DOCUMENTS:
    ** A reference to the applicable design
       documents.
    **
    ** AUTHOR:
    ** Your name here
    **
```

```
**  CREATION DATE:
**  05/14/2005
**
**  NOTES:
**  Other relevant information
*/
```

Note that all the block comments illustrated in this document have no pretty stars on the right side of the block comment. This deliberate choice was made because aligning those pretty stars is a large waste of time and discourages the maintenance of in-line comments.

2.2.3 Revision History

It is sometimes useful to include a history of changes in the source files. With all the new source control tools now available, this information is duplicated in the source control database.

The programmer may use a revision history block in his/her modules. However, only use the revision history block if the programmer intend to maintain it. It is pretty much useless when it's not maintained properly.

Here's the preferred style for a revision history block:

```
/*
**  HISTORY:
**  000 - Nov 91 - M. Taylor      -
```

Glossary

deliberate *adj.* 謹慎的

```
Creation
   ** 001 - Dec 91 - J. Brander    -      Insert validation for
   **                                     unitlength to detect
   **                                     buffer overflow
   */
```

2.2.4 Code Specifications

Documentation comments, specially in Java, describe the classes, interfaces, constructors, methods, and fields. Each documentation comment is set inside the comment delimiters /**...*/, with one comment per class, interface, or member. This comment should appear just before the declaration:

```
/**
 * The Example class provides...
 */
public class Example {...}
```

Notice that top-level classes and interfaces are not indented, while their members are.

If the programmer needs to give information about a class, interface, variable, or method that isn't appropriate for documentation, he/she has to use an implementation block comment (see section 2.1.1) or single-line (see section 2.1.2) comment immediately after the declaration. For example, details about the implementation of a class should go in such an implementation block comment following the class statement, not in the class documentation comment.

Documentation comments should not be positioned inside a method or constructor definition block, because Java associates documentation comments with the first declaration after the comment.

Unit 6 Documenting Your Work

2.2.5 Sources

If the programmer has used a number of lines from some open-source codes, he/she has to document the URL of the code and line numbers he/she took the lines from.

If he/she has used an algorithm from a book or magazine, he/she has to document the volume and page number he/she took it from. If he/she developed the algorithm herself, he/she had better indicate where the reader can find the notes he/she has made about it.

article?

Post-reading exercises

Exercise 1

How many styles of implementation comments are mentioned here? What are they?

Exercise 2

Please connect the following names with their main conventions.

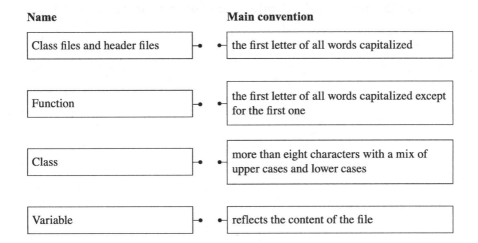

Name	Main convention
Class files and header files	the first letter of all words capitalized
Function	the first letter of all words capitalized except for the first one
Class	more than eight characters with a mix of upper cases and lower cases
Variable	reflects the content of the file

Exercise 3

Translate the following sentences into Chinese.

a) A good name for a routine clearly describes everything the routine does.

b) Special care should be given to the naming of header files because of potential conflicts between modules.

c) When applicable, all source documentation should be in a format compatible with the generic formatting defined by the original programming language.

d) Short comments can appear on a single line indented to the level of the code that follows.

e) Very short comments can appear on the same line as the code they describe, but should be shifted far enough to separate them from the statements.

II E-mail & technical writing
Writing

A Replying or forwarding e-mails properly

E-mails are now considered an important means of communication. In the IT industry, where efficiency and brevity are appreciated, electronic communication has become increasingly popular because of its speed and broadcasting ability. Here is a series of guidelines to help you write effective professional e-mails.

Communicating Professionally and Effectively by E-mail

—Replying & forwarding

√ **Every message other than spam or junk mail deserves a reply.**

To do so takes time, and time is of the essence for everyone. However, notes from people like your boss, your customers, people you care about, and people with whom you have not spoken in a while always merit a reply.

√ **Give your final answer in your reply.**

Professional e-mails usually seek specific answers to questions. When you reply, make sure you answer the question as completely as possible. Do not force the time-pressed message sender to ask the question again, or engage in many exchanges in order to obtain his answer.

√ **Respond as soon as possible.**

If you want to appear professional and courteous, be accessible to your online correspondents. Reply within a couple of hours of receiving the message. Even if your reply is, "Sorry, I'm too busy to help you now," at least your correspondent will not be left waiting for your reply.

√   **When replying, include enough of the original message to provide a context.**

Remember that e-mail is not as immediate as a telephone conversation and the recipient may not recall the contents of the original message, especially if he or she receives many messages each day. Including the relevant section from the original message helps the recipient to place your reply in context.

√   **When replying or forwarding a message, minimize the original one.**

When replying to or forwarding an e-mail, clean up the document. Quote only the smallest amount you need to make your context clear. Use the "BCC" or blind carbon copy command more often than the "CC" or carbon copy command. In the message you forward, delete the extraneous information such as the subject, addresses, and date lines. Doing so pares down the size of the message, making it easier to read. This is just another form of common e-courtesy.

√   **Distinguish between text quoted from the original message and your reply.**

This makes the reply much easier to follow. ">" is a traditional marker for quoted text, but you can use anything provided its purpose is clear and you use it consistently. Whatever you do, make sure your responses are clearly separated from their original text, in order to avoid mixing messages.

√   **Know where your reply or forward will end up.**

It can be embarrassing for you if a personal message ends up on a mailing list, and it is annoying for the other list members.

×   **Reply to e-mails with one-word answers or questions.**

"What?" "What what?" It's another time waster. Avoid it if you can.

×   **Forward e-mails unless you have the permission of the author.**

What they wrote may not have been intended for wider distribution, so it is always wise to ask first. Too often confidential information has gone global because of someone's lack of judgment. Unless you are asked or request permission, do not forward anything that was sent just to you.

× **Forward forwarded messages to your friends and co-workers.**

Chain letters and jokes are junk mail to some. E-mail can be fun, but do not misuse or abuse it.

× **Use "Reply to All" when you are replying to just one sender.**

The "Reply to All" button is just a button, yet it can generate many unnecessary e-mails. Only use "Reply to All" if you really need your message to be seen by each person who received the original message. Make only wise and sparing use of it.

B E-mails about software documentation

1. Gather examples of e-mail replies and forwarded e-mails. Are they properly written in terms of layout, tone, and professionalism? In which aspects do they stand out as excellent? Or, how can they be improved according to our guidelines?

2. Imagine you are replying to an e-mail from a colleague. That colleague is a new employee in the development group, and does not know much about source code documentation, so he/she wrote you a message about copyright comments. You want to reply to him/her noting:

- the importance of commenting on copyright information in source code;
- a brief introduction on how to comment on copyright information;
- that he/she should feel free to reply if there are any other relevant questions.

Send this e-mail to your partner and your instructor, and review the message sent from your partner.

Guide to technical writing

No matter what your current or future job is, writing will be essential to your work because you will have to communicate your technical knowledge to others. Technical documents are a most frequently used type of writing in the IT industry, so training in technical writing will help ensure a smooth career path. Here is a series of guidelines to help you write successful technical documents.

Guidelines to Successful Technical Writing

—Drafting planning proposals

A proposal is an offer to do something or a suggestion for action. The general purpose of a proposal is to persuade readers to improve conditions, authorize work on a project, or support a plan for solving a problem or doing a job.

A planning proposal suggests ways to solve a problem or to bring about improvement. It might be a request for funding to expand the campus newspaper, an architectural design for a new software system, or a plan for researching and developing a whole new operating system. In every case, the successful planning proposal answers this central question for readers:

What are the benefits of following your suggestions or your proposal?

Like any document that gains reader acceptance, it also has to be a product of careful decisions about content, organization, and style. Comply with the following guidelines when you are drafting planning proposals:

Use appropriate format and supplements.

Short plans can take the form of memo, letter or even e-mail, depending on whether they are internal or external. They should be able to serve many of your communication purposes. However, long proposals are often required by some projects. Long proposals usually have headings, appendices, and several other supplements (cover letter, title page, informative abstract, etc.). Therefore, choose an appropriate format tailored to your audience and your objectives.

Write a concrete and specific title.

Begin with a title that is clear about the intent of your proposal:

Vague   A Plan for New Programs at Cosmos Engineering

Concrete A Preliminary Planning Proposal for Developing the New Planet Tracking Software at Cosmos Engineering

Begin with a clear introduction.

In the introduction, answer the *what* and *why*, and clarify the subject, background, and purpose of your planning proposal. Establish need, suggest benefits, and reveal qualifications. Also, reference your data and note any limitations of your proposal.

Develop a plain and understandable body.

In the body of the proposal answer the *how*, *when*, and *how much*. Your proposal should contain methods, work schedules, materials and equipment, personnel, facilities, costs, expected results, and feasibility, and each one should be numerated, i.e. 1, 2, 3, etc.

Design your planning proposal to reflect your attention to detail.

A hastily typed and assembled proposal suggests to readers the writer's careless attitude toward the project. Keep layout, typeface, and bindings conservative and tasteful.

Focus on one specific subject and one purpose.

Readers want specific suggestions for filling specific needs. By spelling out your subject and purpose, show your readers immediately that you understand their problem and want to help them address it.

Identify all problems readers themselves might not recognize.

Do not underestimate the complexity of the proposed project. Remember that only problems that have been fully and clearly defined can possibly be solved.

Provide concrete and specific information.

Vagueness is a fatal flaw in a proposal. Before you can persuade readers, you must inform them; therefore, you need to *show* as well as *tell*. Include graphics essential to making a persuasive case for your proposal.

Use visuals whenever possible.

Use visuals if they enhance your planning proposal. Introduce them and interpret them in the text.

End with a comprehensive summary.

In the conclusion, summarize key points, and stimulate action.

D Technical writing exercises

Exercise 1

Locate a planning proposal from your library or company. Does the proposal conform to the above guidelines in terms of format, content, arrangement, and style? If not, how could the proposal be improved? Discuss your conclusion in class or in writing (one or two paragraphs).

Exercise 2

After consulting with instructors or (other) IT professionals, write a memo to your instructor identifying the kinds of proposals most often written in your field. Are such proposals usually short or long, internal or external? Who are the decision makers in your field? Provide at least one detailed scenario of a work situation that calls for a proposal. Identify both the primary and secondary audience before you begin writing. Include your sources of information in your memo.

Unit 7

Implementing a Project

项目实施

I Project work plans

II E-mail & technical writing

Project work plans
Reading

A An overview

Project Work Plan (PWP)

A project work plan is the schedule of tasks, durations, and dates to accomplish a project. A high level work plan shows the standard deliverables for the process, and a detailed project plan shows the project-specific tasks.

A project work plan provides a plan of what has to be done, how the work will be organized and who will do it. This will require a calendar or chart that provides a schedule of work outlining the project goals, objectives, strategies, and evaluations.

The following steps will provide you with information to be used in developing a good work plan:

- List the project objectives and chosen strategies;
- List the tasks you need to do for those objectives and strategies to be met;
- Put the tasks in the order that they should occur;
- Estimate times and dates to create a schedule;
- Estimate the resources needed for your activities (including time, money, skills, people, equipment, facilities, information, etc.);
- Assign responsibility to people for various activities.

The purpose of the project work plan is to describe how the work of the project will be carried out and the goals of the program accomplished. This includes describing the project management; data collection and management; data quality assessment, validation, and usability and data analysis process.

Unit 7 Implementing a Project

B A sample of project work plan

Read the following project work plan. For the first time, please only scan the whole document. Keep these questions in mind and try to answer them after scanning. Time limit: 10 minutes.

- What is this document mainly talking about?
- How many work packages have been defined?
- Where should we put most of the emphasis in the last 6-12 months of this project?

Speech Recognition Algorithms for Chinese (SPRACH)

Project Work Plan

1. Project Overview

The goal of this national project is to further improve the current technologies and produce an <u>applicable</u> system in continuous Chinese speech recognition. <u>Pursuing</u> the <u>theoretical</u> and design work successfully carried out under several previous theoretical projects, this new project, referred to as SPRACH (Speech Recognition Algorithms for Chinese), will <u>extend</u> the previous artifacts and develop a <u>robust</u> and flexible speech recognition system that can be adapted to new Chinese <u>dialects</u> and new domains. This project is proposed and supervised by General Computers Corporation.

2. Introduction to the Work Plan

In the following list we summarize the work

Glossary

recognition *n.* 识别
applicable *adj.* 切实可行的
pursue *v.* 跟随，继续执行
theoretical *adj.* 理论的
extend *v.* 延续，扩展
robust *adj.* 强壮的
dialect *n.* 方言

packages broken down into their component tasks. In short, on top of Work Package 0 (WP0) on Project Management, five work packages have been defined:

1) WP1: Development of the software and hardware tools necessary to carry out the proposed work. This is particularly important in reducing the research cycle and forcing all the partners to work on the same software and hardware basis;

2) WP2: Database gathering from different sources and set up of baseline systems. In this framework, a large-vocabulary, continuous speech recognizer will be developed for Chinese;

3) WP3: Development of (and development tools for) lexicons, language models (LM), and acoustic models for multiple Chinese dialects;

4) WP4: System implementation and deployment of SPRACH;

5) WP5: Extensive testing, advertising and deployment.

The last 6-12 months of this project will put most of the emphasis on prototype development of WP6 (demonstrations and evaluations).

Strong interaction between all the partners and all work packages will be guaranteed through the use of the same hardware, software and databases.

Only dialect specific developments will be carried out by the respective sites.

> **Glossary**
> lexicon *n.* 词素
> acoustic *adj.* 声学的
> advertising *n.* 广告活动
> deployment *n.* 配置
> emphasis *n.* 侧重点

3. Detailed Work Plan

3.1 Work Package WP0: Management

Work Package Manager: GCC

Executing Partner: THU

See Section 4 for a description of the management structure and follow-up of the different work packages.

Although our research and development work is well defined in the proposal, a kick-off meeting will be organized at a very early stage of the project to guarantee synchronization and supervision and to initiate working contacts with our industrial advisors to understand their concerns and their expectations.

3.1.1 WP0: Milestones

M0.1 (T0+1): Kick-off meeting (with all partners)

M0.x Regular meetings

3.1.2 WP0: Deliverables

D0.x Short management report every 6 months

3.2 Work Package WP1: Hardware and Software Support

Work Package Manager: GCC

Executing Partners: CAS, THU

We will be incorporating new hardware and software from the CAS that provided successful system infrastructures for previous linguistic

> **Glossary**
> kick-off *n.* 开始
> synchronization *n.* 同步
> supervision *n.* 监管
> linguistic *adj.* 语言的, 语言学的

R&D programs with the RAP computers. The new systems, which will be vector-based, will be required in SPRACH to—

- support our research & development (e.g., fast assessment of new research results on moderate sized tasks);
- provide fast training on large databases.

Towards these ends the CAS team has developed a vector supercomputer on a chip called the Dragon Heart. Software is currently being developed for this chip that is analogous to the routines currently used on the RAP. Meanwhile, the CAS team is finalizing the building of a single-processor S-bus board called Phoenix that will be made available to the SPRACH partners for their research and development work. During SPRACH, the THU team will develop multi-node versions called SPRACHStation that will also be made available to the SPRACH partners. The Phoenix boards, which we estimate to be several times faster than the partners' RAP systems, while being a small fraction of the price, are expected to be the main workhorses for the partners. The multi-node SPRACHStation, however, is likely to be required to keep up with new computationally demanding R&D directions and for the large-vocabulary systems.

3.2.1 Task T1.1: Hardware

Task Coordinator: GCC

Executing Partner: CAS

Since the Dragon Heart and Phoenix hardware are produced for another national computational

Glossary

meanwhile *adv.* 同时
phoenix *n.* 凤凰
workhorse *n.* 高负荷机器
computationally *adv.* 计算上地

linguistic project and will be finished soon, no funding will be required for this task. The main effort of this WP will be related to the development of the software tools as well as software adaptation required by this project (Task T6.2).

3.2.2 Task T1.2: Software

Task Coordinator: GCC

Executing Partner: THU

The THU team will be building up the software building blocks necessary for the partners to develop research software easily. In particular, they will provide feedforward and recurrent network modules that take IEEE floating point as input and provide IEEE floating point at the output, while doing arithmetic internally in fixed point for the sake of speed.

3.2.3 WP1: Milestones

- M1.1　(T0+3): Phoenix hardware available on each site.
- M1.2　(T0+3): First version of the SPRACH-Station software available on each site.
- M1.3　(T0+12): First version of training system on Phoenix.

3.2.4 WP1: Deliverables

- D1.1　(T0+3): Report on Phoenix hardware and SPRACHStation software.
- D1.2　(T0+6): Training available on Phoenix.

> **Glossary**
>
> adaptation *n.* 改编, 改写
> recurrent *adj.* 循环的, 递归的
> arithmetic *n.* 算术

D1.3　　　(T0+8): Final training system on Phoenix.

3.3 Work Package WP2: Databases and Baseline Systems

Work Package Manager: GCC

Executing Partners: THU, CAS, CASS

Two essential objectives of the SPRACH project are to

- develop forefront development tools and applications in large-vocabulary continuous speech recognition of the Chinese language;
- conduct state-of-the-art research on the multi-dialectic aspects of Chinese recognition.

This work package has three essential objectives:

- To develop baseline recognizers for the target language in this project—Chinese (Task T1.1);
- To follow the constant evolution of the internationally accepted databases for evaluation of continuous speech recognizers with very large vocabularies (addressed in Task T1.2);
- To provide the databases that are needed for the multi-dialect aspects of the project, namely for Mandarin, Cantonese and other Chinese dialects (also addressed in Task T1.3).

Glossary

essential *adj.* 关键的
forefront *n.* 最前线
internationally *adv.* 国际性地

3.3.1 Task T2.1: Baseline System for Mandarin

Task Coordinator: GCC

Executing Partner: THU

Starting from the experiences in developing large-vocabulary continuous speech recognizer for other languages (e.g. US English), the early task of GCC will be to develop an equivalent system in and for Chinese. This will involve:

- Research and develop applicable algorithms for Chinese linguistic corpora;
- Developing the infrastructure of the recognizer;
- Accessing and processing the Mandarin database available at THU;
- Obtaining or generating (if not available) phonetic transcriptions of the lexicon and the language model;
- Training acoustic models;
- Assessment and comparison with existing systems.

The system resulting of this task will be used:

- to assess and compare with other Mandarin recognizers;
- as a baseline system for dialect specific (Mandarin) developments.

3.3.2 Task T2.2: Baseline System for Cantonese and Other Dialects

Task Coordinator: GCC

> **Glossary**
>
> corpus *n.* (pl. corpora) 语料库
> phonetic *adj.* 语音的
> transcription *n.* 转录,翻译

Executing Partners: CASS, THU

As noted earlier, there is no adequate database for speaker-independent, large-vocabulary, continuous speech recognition in Cantonese and other dialects at present. Such a database will be collected from another small computational project conducted by the CASS. Consequently, a baseline system for Cantonese and other dialects will have to be based on a much smaller database, in order to be available in useful time.

The above mentioned database is not labeled, and there is no adequate labeled database for our target dialects. Therefore, the development of this baseline system will consist of the following three steps:

1) Automatic labeling of the available database, using an application developed by THU.

2) Training the acoustic models based on the labeling from Step 1.

3) Re-labeling of the database using the acoustic models from Step 2. Steps 2 and 3 may have to be iterated several times, to improve the acoustic models and the labeling.

This procedure will use phonetic transcriptions adapted from the available database (which is sentence-based, instead of word-based), and a language model extracted from the Mandarin corpus.

The performance of the baseline system will

Glossary

consequently *adv.* 因而, 所以

label *v.* 分类, 标注

be assessed, but it cannot be expected to equal those of recognizers for Mandarin or other languages, based on much larger speech corpora.

3.3.3 WP2: Milestones

- M2.1 (T0+3): Mandarin database adapted from THU
- M2.2 (T0+15): Baseline Mandarin recognizer available
- M2.3 (T0+24): Database of Cantonese and other dialects available
- M2.4 (T0+24): Baseline recognizer for Cantonese and other dialects available

3.3.4 WP2: Deliverables

- D2.1 (T0+15): Baseline Mandarin recognizer
- D2.2 (T0+24): Baseline recognizer for Cantonese and other dialects

3.4 Work Package WP3: Lexicons, Language Models and Acoustic Models

Work Package Manager: CASS

Executing Partners: THU, GCC

The goals of large-vocabulary, speaker-independent recognition, domain adaptation and task independence require availability of appropriate pronunciation lexicons, and efficient representation and use of language models (LM) and acoustic models. Owing to the diversity in speech between various Chinese dialects, we have to deal with them

> **Glossary**
>
> pronunciation *n.* 发音
> diversity *n.* 多样性

separately. One group is the Mandarin Dialects, which includes a number of dialects considered variants of Mandarin. For this group, there already is a significant amount of speech recognition work done, and consequently there already exist relatively large lexicons, LM and acoustic models, which may however have to be augmented. For the other group, the situation is that of an almost virgin work field, in terms of speech recognition. Development of dictionaries, LMs, and acoustic models for the two groups will, therefore, have to follow different strategies, as outlined in the descriptions of the following tasks.

3.4.1 Task T3.1: Dictionaries and Automatic Learning

Task Coordinator: THU

Executing Partners: GCC, CASS

The two new groups explicitly considered in this project are Mandarin Dialects and the other dialects.

Regarding the first group, the main goal of this task is just to get access to one of the existing large Mandarin Dialect dictionaries. The obvious candidate will be the MDD corpus in THU.

For the second group the situation is however different. As said above, there currently exist no pronunciation dictionaries that can be used for speech recognition. A baseline dictionary will be adapted from the phonetic transcriptions by CASS

> **Glossary**
>
> variant *n.* 变体
> augment *v.* 增加，补充
> strategy *n.* 策略
> explicitly *adv.* 明白地，明确地

(which are based on sentences, instead of words). This will encompass about 2000 words. This dictionary will expand by automatic means during this task. A larger dictionary will be made by hand only if insurmountable difficulties are encountered. With automatic learning ability of new lexicons, we will further investigate a number of ways to derive multiple pronunciations for words.

3.4.2 Task T3.2: Language Model Adaptation

Task Coordinator: CASS

Executing Partner: THU

An appropriate language model is an invaluable component to any speech recognition system larger than a few isolated words. When dealing with a new dialect, or with a dialect whose characteristics change slowly in time, it would be very useful to build on previous dialect models, instead of re-building the language model from scratch. This would have the main advantages of requiring much fewer data, possibly being done in an unsupervised way, and also of saving processing time.

- Adaptation of language models: starting from a given language model (Mandarin LM), adapting it quickly on a new domain, using only a few examples. This can be done in a supervised or unsupervised way. We will investigate methods to use changes in lower-order statistics to modify, in a useful way, higher-order statistics

> **Glossary**
>
> encompass *v.* 包含, 包括
>
> insurmountable *adj.* 不可克服的
>
> investigate *v.* 研究, 试验
>
> from scratch 从零开始

(e.g. trigram probabilities). Lower-order statistics can be obtained with less data, and can therefore yield a quicker adaptation of the language model.

- Dynamic language models: The subject of a conversation is quasi-stationary but this fact is not exploited in current language models. Crude implementations, such as cache-based language models show a modest improvement in perplexity though this has yet to be transferred to increased recognition rates. We propose a clustering scheme to train many independent dialect models and to adaptively mix these based on the past decoded utterances.

3.4.3 Task T3.3: Training Independent Tasks

Task Coordinator: CASS

Executing Partners: THU, GCC

For task independency we wish our acoustic models to be independent of the acoustic environment and independent of the lexicon. Currently our base system is trained for one acoustic condition—that of noise-free read speech with a known microphone. Thus the issues are:

- Incorporating noise robust acoustic vectors;
- Compensating the system for channel variations;
- Compensating the system for known noise conditions;

Glossary

yield *v.* 产生

quasi-stationary *adj.* 准稳定的

exploit *v.* 利用

perplexity *n.* 复杂性

clustering *n.* 集群，聚类

scheme *n.* 计划

utterance *n.* 发音，发声

compensate *v.* 弥补

- Comparison of our baseline context-independent phoneme models with context-dependent HMM/ANN phoneme models to assess the impact on task dependency.

3.4.4 WP3: Milestones

M3.1   (T0+12): Baseline lexicons available

M3.2   (T0+24): Preliminary tests of different language models (LMs) on baseline system

M3.3   (T0+36): Incorporation of LM adaptation in baseline systems

3.4.5 WP3: Deliverables

D3.1   (T0+12): Software for importing baseline dictionary and automatic learning of lexicons

D3.2   (T0+36): Report on LM adaptation

3.5 Work Package WP4: System Implementation and Deployment

Work Package Manager: GCC

Executing Partners: THU, CAS, CASS

After a final compilation, the system will be packaged into a distributable application suite available to run in other R&D sites. Based on the detailed design of the system, General Computers Corporation will take charge of this implementation process. The THU team will help in developing an acoustic decoder.

> **Glossary**
>
> phoneme *n.* 音位, 音素
> compilation *n.* 汇编

3.5.1 Task T4.1: Implementation

Task Coordinator: GCC

Executing Partners: THU, CAS, CASS

The GCC speech recognition engine developed by the GCC Corporation offers efficient decoding, together with a simple interface to language models of arbitrary complexity. Based on this engine and the huge amount of linguistic data gathered in the previous WPs, a brand new Chinese speech recognition system will be developed. This system will be creative algorithmically and implementationally.

The algorithmic developments will include the investigation of fast look-ahead techniques to further prune the search space, and the comparison and optimization of different search strategies. The current algorithm uses a combination of best-first and breadth-first searches. A depth-first search, ordered by language model information, will produce a significant speedup. We plan to investigate the performance of this less-time-synchronous style of search.

Implementational improvements will include improvements in memory and CPU efficiency, with the aim of achieving a single workstation demonstration system that consistently runs in real-time or less. Other improvements include methods to deal with very large vocabularies (100,000 words and larger). The basic search strategy

Glossary

arbitrary *adj.* 任意的

prune *v.* 删减, 削除

employed in GCC speech recognition engine makes it possible to deal with large vocabularies easily, since not all dictionary words need to be present in the language model. This means that vocabulary size may be increased merely by adding to the pronunciation dictionary, without recomputing the language model.

3.5.2 Task T4.2: Deployment

Task Coordinator: GCC

Executing Partners: THU, CAS, CASS

The system will first be deployed in THU, CAS and CASS. These three sites will help in beta testing of the complex speech recognition system. This also gives way to the trail of the compatibility and availability of the system through an extensive deploying process.

3.5.3 WP4: Milestones

M4.1 (T0+24): First version of large-vocabulary recognition of continuous input for Chinese available

M4.2 (T0+36): Real time large-vocabulary recognition of continuous input for Chinese available

M4.3 (T0+60): Limited deployment successful

3.5.4 WP4: Deliverables

D4.1 (T0+24): Large-vocabulary recognition of continuous input for Chinese

D4.2 (T0+36): Real time large-vocabulary recognition of continuous input for Chinese

3.6 Work Package WP5: Evaluations and Prototypes Development

Work Package Managers: GCC, THU

Executing Partner: CAS

In recent years formal evaluations have provided a way of both testing and

quantifying the effect of algorithmic developments in speech recognition and disseminating these advances to the research community. Several partner universities, academies and institutes will give the system an evaluation.

3.6.1 Task T5.1: Evaluations

Task Coordinator: GCC

Executing Partner: CAS

To evaluate the progress of the project quantitatively, we plan to take part in the ACME evaluation program in the US. We have benefited greatly from its supply of resources such as acoustic and language model training data.

3.6.2 Task T5.2: Demonstrations

Task Coordinator: GCC

Executing Partners: THU, CAS

We will release demos to a limited number of public testers to ensure the integrity of the system.

The following prototype systems will be developed and demonstrated during the present project:

- Large Mandarin vocabulary, continuous read speech recognition;
- Small, fast and robust small vocabulary recognition;
- Multi-dialect recognizers.

3.6.3 WP5: Milestones

M5.1   (T0+4): SPRACH system evaluated

Glossary

quantify *v.* 量化

quantitatively *adv.* 定量地

Unit 7 Implementing a Project

M5.2   (T0+8): Production and distribution of all demonstrations

3.6.4 WP5: Deliverables

D5.1   (T0+0): Real-time large-voca-bulary recognition of continuous input for Chinese

D5.2   (T0+8): Three sets of demos

4. Milestones

SUMMARY OF MILESTONES		
Milestone	Deadline (Week)	Description
WP0: Management		
M0.1	1	Kick-off meeting
+ Regular meetings as planned		
WP1: Hardware and Software Support		
M1.1	3	Phoenix hardware available on each site
M1.2	3	First version of the SPRACHStation software available on each site
M1.3	12	First version of training system on Phoenix
WP2: Databases and Baseline Systems		
M2.1	3	Mandarin database adapted from THU
M2.2	15	Baseline Mandarin recognizer available
M2.3	24	Database of Cantonese and other dialects available
M2.4	24	Baseline recognizer for Cantonese and other dialects available
WP3: Lexicons, Language Models and Acoustic Models		
M3.1	12	Baseline lexicons available
M3.2	24	Preliminary tests of different language models (LMs) on the baseline system
M3.3	36	Incorporation of LM adaptation in baseline systems
WP4: System Implementation and Deployment		
M4.1	24	First version of large-vocabulary recognition of continuous input for Chinese available

(Continued)

M4.2	36	Real-time large-vocabulary recognition of continuous input for Chinese available
M4.3	60	Limited deployment successful
WP5: Evaluations and Prototypes Development		
M5.1	4	SPRACH system evaluated
M5.1	8	Production and distribution of all demonstrations

Post-reading exercises

Exercise 1

How many milestones are there in WP2? What are they? What are the deadlines (in weeks) of each milestone?

Exercise 2

Please connect the following work packages with their primary tasks.

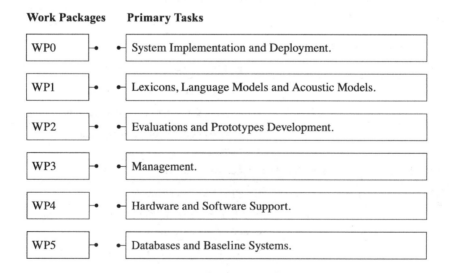

Exercise 3

Translate the following sentences into Chinese.

a) Strong interaction between all the partners and all work packages will be guaranteed through the use of the same hardware, software and databases.

_____

_____

b) As noted earlier, there is no adequate database for speaker-independent, large-vocabulary, continuous speech recognition in Cantonese and other dialects at present.

_____

_____

c) Regarding the first group, the main goal of this task is just to get access to one of the existing large Mandarin Dialect dictionaries.

_____

_____

d) As said above, there currently exist no pronunciation dictionaries that can be used for speech recognition.

_____

_____

e) A larger dictionary will be made by hand only if insurmountable difficulties are encountered.

_____

_____

11 E-mail & technical writing
Writing

A Using abbreviations and smileys

E-mails are now considered an important means of communication. In the IT industry, where efficiency and brevity are appreciated, electronic communication has become increasingly popular because of its speed and broadcasting ability. Here is a series of guidelines to help you write effective professional e-mails.

Communicating Professionally and Effectively by E-mail

—Abbreviations & smileys

√ **Learn the most common e-mail abbreviations.**

Abbreviation usage is quite widespread in e-mail communications, so acquaint yourself with them. Some of the most common are listed in the table below. It is recommended that you use abbreviations already common to the English language, such as FYI and BTW. Beyond that, you run the risk of confusing your recipient.

Abbreviation	Meaning
AFAIK	as far as I know
AKA	also know as
ASAP	as soon as possible
B4	before
BBL	be back later
BCNU	be seeing you
BFN	bye for now

(Continued)

Abbreviation	Meaning
BTW	by the way
CU	see you
CUL	see you later
FWIW	for what it's worth
FYA	for your amusement
FYI	for your information
GMTA	great minds think alike
HTH	hope this (that) helps
IAE	in any event
ICWUM	I see what you mean
IMO	in my opinion
IMHO	in my humble opinion
IOW	in other words
J4F	just for fun
LOL	laugh out loud
TTYL	talk to you later

√ **Make sure the recipients understand the abbreviations.**

If you use abbreviations or acronyms, be sure your audience already knows what they stand for. If you are unsure that they do, use the abbreviation with an explanation immediately behind it. You want your readers to feel comfortable.

√ **Learn the most common smileys.**

E-mail closely matches the immediacy of a conversation, but is devoid of "body language". Users have come up with something called smileys, or emoticons for usage in e-mails to try and fill that void. They are simple strings of characters that are interspersed in the e-mail text designed to convey the writer's emotions (body language). The most common example is :-). Turn your head to the left and you should see a happy face (the colon are the eyes, the dash is the nose and the parentheses is the mouth). Here are some more examples with their possible meanings.

Smiley	Possible Meaning
:-) *or* :)	Smiling face
;-) *or* ;)	Wink
:-l *or* : l	Indifference or embarrassment
:-> *or* : >	Heavy grin
8-) *or* 8)	Eye-glasses
:-D *or* : D	Laugh, shock or surprise
:-/ *or* : /	Perplexed
:-(*or* : (	Frown (anger or displeasure)
:-P *or* : P	Wry smile
;-} *or* ; }	Leer
:-Q *or* : Q	Smoker
:-e *or* : e	Disappointment
:-@ *or* : @	Scream
:-O *or* : O	Yell
:-* *or* : *	Drunk
:-{} *or* : {}	Wearing lipstick or surprised

√ **Use smileys to avoid misunderstandings.**

Using the common smileys carefully can markedly improve the clarity of your message, since they convey nuances which approximate body language. You may think you are being funny (or serious) in your writing, but it may appear differently to the reader. If you want to ensure their understanding, use a smiley or two. There are thousands of possible combinations. If you're trying to be funny, inserting a smiley face can convey your intent, preventing misunderstandings or hurt feelings. Likewise, if your writing is not indicative of your foul mood, perhaps a frowning emoticon should be inserted at the end of the sentence.

× **Overuse abbreviations that are likely to cause frustration.**

PC users have their own shorthand language that uses expressions designed to save typing, such as IMHO (in my humble opinion) and TTYL (talk to you later). However, many people find these abbreviations as unintelligible as organic chemistry formulas. You can not assume everyone is familiar with the endless acronyms circulating out there. WIDLTO—when in doubt, leave them out.

× **Overuse emoticons that damage your image.**

Like any embellishment, however, overuse of smiley faces destroys their value. So use them sparingly.

B Exercises on e-mail writing

1. Gather sample e-mails from corporate offices and from your personal home files. Compare these two types of e-mails. Tally how frequently abbreviations and smileys are used in them. Are there any differences with and without abbreviations and smileys? Which style do you prefer? Write an e-mail to your instructor about your thoughts.

2. You are writing to a client about a change in your schedule. You had planned to have dinner with her next Friday to talk about a big contract, but had to postpone it. You want to:

- Apologize about the change in the dinner appointment, but you have to be on a business trip to India;
- Ask if it is possible for her to alter her schedule so you can meet at the beginning of the next month;
- Add that if she can not decide now, you two can discuss it after you are back from India;
- Emphasize how much you hope for future cooperation with her company.

Send this e-mail to your partner and your instructor, and review the e-mail message from your partner.

Guide to technical writing

No matter what your current or future job is, writing will be essential to your work because you will have to communicate your technical knowledge to others. Technical documents are a most frequently used type of writing in the IT industry, so training in technical writing will help ensure a smooth career path. Here is a series of guidelines to help you write successful technical documents.

Guidelines to Successful Technical Writing

—Revising with efficient sentences

The main purpose of a technical document is to inform or persuade the reader through use of efficient sentences, not to impress or entertain with fancy language displays. So, technical documents transmit worthwhile information—even highly specialized information—in the most straightforward way to their audience.

Readers of technical documents are busy and impatient. They do not wish to put more into reading a document than they can get from it. They hate waste and expect efficiency. Every sentence in a document should be revised to carry its own weight, in conveying the writer's meaning.

Observe the same rule in adopting sentence style as you would in choosing the document's content: make it long enough to be understood, yet short enough to be tolerated. When writing a technical document, consult the guidelines below, which can also serve as a checklist for your sentence style:

Revise sentences to be clear and avoid ambiguity.

A clear sentence communicates its precise meaning on first reading. It signals relationships among its parts, and emphasizes the key thought. In technical writing, a sentence should have only one meaning. Make sure the words, phrases, and punctuation are absolutely clear.

Ambiguous　I cannot recommend this candidate too highly.

Revised　　This candidate has my highest recommendation.

Unit 7 Implementing a Project

Ambiguous Many executives are skeptical about office automation as well as managers.

Revised Many executives as well as managers are skeptical about office automation.

Ambiguous Being well-known in the computer industry, our project would benefit a lot from the Tsinghua team.

Revised Because the Tsinghua team is well-known in the computer industry, our project would benefit a lot from its help.

Use active rather than passive voice most of the time.

Usually, the active voice (Joe tested the software.) is better than the passive voice (The software is tested by Joe.), but in certain situations it can make sense to use the less natural passive style. However, many writers routinely use the passive style simply because they believe it is more "formal" and "acceptable". It is not. Using the passive style is the most common reason for poorly structured sentences and it always leads to longer sentences than are necessary. Unless you have a very good reason for the change in emphasis, you should always write in the active style.

The following examples show the improvement achieved by switching from passive to active:

Faulty   The report was written by Peter, and was found to be excellent.

Correct   Peter wrote the report, and it was excellent.

Faulty   The lid should be sealed with wax.

Correct   Seal the lid with wax.

Bad   The values were measured automatically by the control system.

Good   The control system measured the values automatically.

Weak & Impersonal   An offer will be made by us next week.

Stronger   We will make you an offer next week.

Avoid unnecessary words and repetition.

Many sentences contain unnecessary words that repeat an idea already expressed in another word. This wastes space and blunts the message. Often writers use several words for ideas that can be expressed in one. This leads to unnecessarily complex sentences and redundancy.

Redundant   The printer is located adjacent to the computer.

Revised   The printer is adjacent to the computer.

Redundant   The user can visibly see the image moving.

Revised   The user can see the image moving.

Redundant   The product is not of a satisfactory nature.

Revised   The product is unsatisfactory.

Make sentences fluent.

Fluent sentences are polished, graceful, and easy to read. Varied length and word order make them free of choppiness and monotony. See the following suggestions:

Needless "that"   This is a problem that bothers me.

Fluent   This problem bothers me.

Needless qualifier   It seems that they have a valid argument.

Revised   They have a valid argument. / They seem to have a valid argument.

Choppy   Jogging can be healthful if you have the right equipment. Most important are well-fitting shoes. They are important because without them you take the chance of injuring your legs. your knees are especially prone to injury.

Revised   Jogging can be healthful if you have the right equipment. Well-fitting shoes are most important because they prevent injuries to your legs, especially your knees.

Unit 7 Implementing a Project

D Technical writing exercises

Exercise 1

The following sentences are unclear or lack fluency because of ambiguous phrasing, incorrect word order, or too much information. Revise them so that their meanings are clear.

  a) A man eating shark was spotted in the South Sea.

  b) Wearing special equipment, the radioactive material failed to injure the operator.

  c) That is a whole new approach that needs attention and research.

Exercise 2

The following sentences need to be rewritten in the active or passive voice for better emphasis, more directness or greater economy. Make necessary changes and give reasons for each.

  a) It is believed by us that the contract is faulty.

  b) Special helmets should be worn at all times during this project.

  c) It was reported by the manager that the project was in trouble.

  d) Care should be taken in the operation of the machine.

  e) The stability of the process is enhanced by co-operation.

Exercise 3

Fill in the following weekly status report, showing your progress toward completing a recent software project. Or, organize into groups of four, and create on paper a scenario in which your group is developing a database for Southwest Airlines. Or, compose a short description of your progress in learning technical writing.

Project Demographics	Values
Identifier (organization-specific code)	
Name of Project	
Customer Name/Organization	
Primary Customer Interface Person	
Project Start Date	
Project Finish Date	
Total Projected Work Effort (Person-days)	
Phase of Project/Current Milestone	
Key Roles in Project	**Names of People in the Roles**

Project Objectives

Customer Objectives

Provider Objectives

Unit 7 Implementing a Project

Deliverables to be Provided	Review Process	Completion Date

Milestones	Baseline Target Date	Actual Result Date	Estimated Effort	Actual Effort

Unit 8

Negotiating Assignments

协商任务

I   Unit test plans

II   E-mail & technical writing

Unit test plans
Reading

A An overview

Unit Test Plan

A unit test plan is a document describing the unit testing process in terms of the features to be tested, pass/fail criteria and testing approaches, resource requirements and schedules.

Unit testing is the process of testing the individual sub-programs, sub-routines or procedures in a program.

Unit testing involves testing the smallest possible unit of an application. Unit testing is recognized as an essential component of the software development process. Unit testing practitioners enjoy such benefits as easier error detection, which has the very desirable end result of increasing software quality while reducing development time and cost. Easier error reduction leads to reduced development time, effort, and cost, because less time and resources are consumed in finding and fixing errors. In addition, unit testing involves several complex types of testing:

- White box testing: ensures that code is constructed properly and does not contain any hidden weaknesses.
- Black box testing: ensures that code functions in the way that it is intended to function.
- Regression testing: ensures that modifications do not introduce errors into previously correct code.

B A sample of unit test plan

Read the following unit test plan. For the first time, please only scan the whole document. Keep these questions in mind and try to answer them after scanning. Time limit: 10 minutes.

- What is this document mainly talking about?
- Why should we do unit tests according to this Unit Test Plan?
- What different tasks are necessary for preparing and performing a specific unit test?

Interface Specification Tool

for Speech Recognition Algorithms for Chinese Project at General Computers Corporation

Unit Test Plan

VERSION 0.0.5

A. Abstract

This document describes the Unit Test Plan (UTP) for the project and is written according to the software engineering standard provided by the General Computers Corporation. The project's unit test for the product is described here. This project is part of the SPRACH Project and defines interface functions for the whole system.

B. Table of Contents

- Abstract
- Table of Contents
- Document Status Sheet
- Document Change Record

1. Introduction
   1.1 Purpose
   1.2 Scope
   1.3 List of Definitions
   1.4 List of References
2. Test Plan
   2.1 Test Items
   2.2 Features to be Tested
   2.3 Test Deliverables
   2.4 Testing Tasks
   2.5 Environmental Needs
   2.6 Test Case Pass/Fail Criteria
3. Test Case Specifications
   3.1 Model Data Storage
   3.2 Undo Assistant
   3.3 Model Verifier
   3.4 Event Router
   3.5 GUI
   3.6 TGUID
   3.7 TXMLParser
4. Test Reports
   4.1 Model Data Storage
   4.2 Undo Assistant
   4.3 Model Verifier
   4.4 Event Router
   4.5 GUI

4.6 TGUID

4.7 TXMLParser

Document Title	Unit Test Plan
Author(s)	Sam Carter, Frank Jefferson
Version	0.0.5
Document Status	draft/internally accepted/conditionally approved/approved

Version	Date	Author(s)	Summary
0.0.1	25-03-20xx	Sam Carter, Frank Jefferson	Document creation.
0.0.2	19-05-20xx	Sam Carter, Frank Jefferson	Revised document after specifying tests, added test specifications.
0.0.3	21-05-20xx	Sam Carter, Frank Jefferson	Specific tests added to Chapter 3.
0.0.4	27-05-20xx	Sam Carter, Frank Jefferson	Test reports added to Chapter 4.
0.0.5	03-06-20xx	Sam Carter, Frank Jefferson	More test reports added to chapter 4, chapter 3 revised.

D. Document Change Record

Document Title	Unit Test Plan
Date of Changes	03-06-20xx

Section Number	Reason for change
3	Revised some tests.
4	Test reports added.

1. Introduction

1.1 Purpose

This document describes the plans for testing the developed software units as defined in the Architectural Design Document (ADD) and the Detailed Design Document (DDD). The purpose of the unit tests is to make sure that the software units, or code modules, developed by the Interface Specification Tool project team comply with the requirements as stated in Chapter 4 of the ADD and Chapter 3 of the DDD. These tests must be executed in the detailed design phase of the project.

1.2 Scope

In the second chapter the items to be tested and all things needed for the tests are mentioned. The type of assignment makes it difficult to describe the test case specifications and the test procedures in two chapters, so the cases and procedures are "combined" into one chapter (chapter 3). The fourth chapter presents the reports for all test cases.

1.3 List of Definitions

ADD	Architectural Design Document
DD	Detailed Design
DDD	Detailed Design Document
GUI	Graphical User Interface
IST	Interface Specification Tool
ITP	Integration Test Plan
SCMP	Software Configuration Management Plan
SPRACH	Speech Recognition Algorithms for Chinese
URD	User Requirements Document
XML	EXtensible Markup Language

Glossary

assignment *n.* 任务

1.4 List of references

ADD	Architectural Design Document, Susan Thomson, Jane Baker, Wang Li, General Computers Corporation, 20xx
DDD	Detailed Design Document, Susan Thomson, Jane Baker, Wang Li, General Computers Corporation, 20xx
ISPEC	ISpec: Towards Practical and Sound Interface Specifications (IFM 2000, LNCS 1945, pp. 116-135, Springer-Verlag), Hans B.M. Jonkers, Philips Research Laboratories Eindhoven, 2000
ITP	Integration Test Plan, Susan Thomson, Jane Baker, Wang Li, General Computers Corporation, 20xx
SCMP	Software Configuration Management Plan, David Zhang, General Computers Corporation, 20xx
URD	User Requirements Document, John Smith, Liu Feng, Michael James, General Computers Corporation, 20xx
UTP	Unit Test Plan, Sam Carter and Frank Jefferson, General Computers Corporation, 20xx

2. Test Plan

2.1 Test Items

The software to be tested consists of all the code modules developed for the IST Project. In this UTP, the difference is made between low-level code modules and high-level code modules. A low-level code module normally is a single class, which is saved as a source file. The classes are defined in the ADD and the DDD.

A high-level code module is represented by a component. Components are composed of one or more low-level code modules and are described in the ADD and DDD.

2.2 Features to Be Tested

All code modules to be tested are listed here. Their requirements are listed in the ADD and the DDD.

In the following table, the high-level code modules to be tested are listed, together with their code module identifier names and the paragraphs where their test cases, procedures and reports are described.

High-level code module	Paragraphs	
Model DataStorage	3.1	4.1
Undo Assistant	3.2	4.2
Model Verifier	3.3	4.3
Event Router	3.4	4.4
GUI	3.5	4.5

In the following table, the low-level code modules to be tested are listed, together with the paragraphs where their test cases, procedures and reports are described in this document. The code module identifier names, which are used for the unit tests, are exactly the same as the name of the low-level code modules.

Code module	Paragraphs	
TGUID	3.6	4.6
TXMLParser	3.7	4.7

The correspondence between high-level and low-level code modules is listed below.

High-level code module	Corresponding low-level code modules
Model DataStorage	TGUID, TXMLParser

There are a few remarks to this:

1) Before a high-level code module is to be unit tested, all corresponding low-level code

Glossary

correspondence *n.* 对应

Unit 8 Negotiating Assignments

modules have to be unit tested;

2) When a high-level code module is unit tested, it means that the low-level code modules are unit tested and that the integration between these low-level code modules is tested. Because the integration between low-level code modules is, if applicable, so fine-grained and simple of nature, it is put into the unit test of the high-level code module. This way superfluous integration tests are prevented and the Integration Test Plan (ITP) is more efficient.

2.3 Test Deliverables

The following items must be delivered before being able to test at all:

- The Architectural Design Document;
- Chapters 1 and 2 of the Detailed Design Document;
- Chapters 1 and 2 of this Unit Test Plan.

The following items must be delivered before a specific unit test begins:

- Paragraphs concerning the code module to be tested in Chapters 3 of the DDD;
- Paragraphs concerning this specific unit test in Chapter 3 and 4 of this document;
- Code module to be tested;
- Test harness for this specific unit test (this is described below);
- Input test data for this specific unit test.

A test harness is a collection of drivers and stubs.

> **Glossary**
>
> integration *n.* 集成
> fine-grained *adj.* 平滑的
> superfluous *adj.* 多余的
> prevent *v.* 防止
> harness *n.* 套具
> stub *n.* 存根

A driver is a main program that accepts test data and passes the data to the module to be tested and prints relevant results. A stub simulates a subordinate module that is called by the module to be tested.

The test harness, i.e. the drivers and the stubs, must be kept for future re-run of tests. The storage of these files is described in Appendix B.3 of the Software Configuration Management Plan (SCMP).

The following items must be delivered when a specific unit test is finished:

- Unit test report, which will be listed in the unit test's paragraph in Chapter 5 of this document;
- Unit test output data;
- Bug reports (if necessary).

The following items must be delivered when testing on all code modules has finished:

- All unit test reports, which comprises Chapter 5 of this document.

2.4 Testing Tasks

The following tasks are necessary for preparing and performing a specific unit test:

- designing the unit test;
- designing a test harness;
- designing input test data;
- setting up a system, which conforms the environmental needs of Chapter 2.5 with the specific unit test's environmental

Glossary

subordinate *adj.* 从属的
comprise *v.* 包含
conform *v.* 使符合

Unit 8 Negotiating Assignments

needs, the unit to be tested, the test harness and the input test data;
- performing the unit test.

2.5 Environmental Needs

- A computer with at least the following specifications: 192 MB main memory, Inter Pentium III equivalent processor or better, 250 MB of free hard disc space (excluding Visio);
- Microsoft Windows 2000 or Windows XP;
- Microsoft Visual Basic 6.0 SP5;
- vbUnit 3.06.02 Evaluation version, a "framework" to support unit testing. More info about it can be found at www.vbunit.org, and in CVS (./DevelopmentLibrary/Miscellaneous/Tools);
- Microsoft Visio 2002 professional SR-1;
- The used system must have enough free secondary store available to perform the test case.

If a test case needs specific environmental needs, this will be mentioned in Chapter 3 in the table entry "Environmental needs".

2.6 Test Case Pass/Fail Criteria

Every test case must describe what the criteria are to pass that specific test.

3. Test Case Specifications

3.1 Model Data Storage

Test Case Identifier	ModelDataStorageT1.
Test Item(s)	IModelDataStorage.Initialize, IModelDataStorage.Deref
Input Specification	initialize(parVisio, parIEventRegistration) is called once. deref() is called once.
Output Specification	No exceptions are raised.
Environmental Needs	A stub parVisio that implements the IVisio interface. A stub parEventRegistration that implements the IEventRegistration interface.

In some of the test cases Figure 1 will be used:

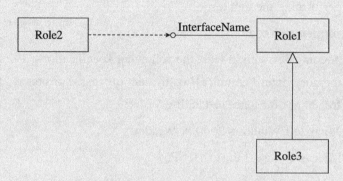

Figure 1 Model Used for testing of MDS

(Other test cases are removed by the editor)

3.2 Undo Assistant

Test Case Identifier	UndoAssistantT1
Test Item(s)	IUndoAssistant.initialize, IUndoAssistant.deref
Input Specification	initialize(parStorageAlter, parStorageRetrieve, parStorageRegistration, parVisio) is called once. deref() is called once after initialize
Output Specification	No exceptions are raised.
Environmental Needs	A stub parStorageAlter that implements the IStorageAlter interface. A stub parStorageRetrieve that implements the IStorageRetrieve interface. A stub parStorageRegistration that implements the IStorageRegistration interface. A stub parVisio that implements the IVisio interface.

(Other test cases are removed by the editor)

- The Undo Assistant is an observer of MDS. It should create undo information.

Unit 8 Negotiating Assignments

- The Undo Assistant should receive <u>notifications</u> when an undo action is started.
- Try to pass <u>erroneous</u> information with the change notifier.

Glossary

notification *n.* 通知

erroneous *adj.* 错误的

3.3 Model Verifier

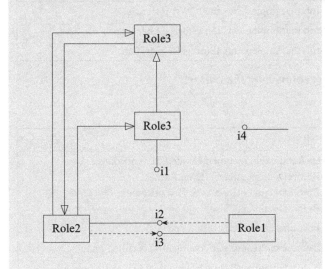

Figure 2 An example of a model

Test Case Identifier	ModelVerifierT1
Test Item(s)	ICWA.getSatisfiesCWA, ICWA.getCWAResInterfaces1(), ICWA.getCWAResInterfaces2(), ICWA.getSatisfiesCWAResRoleCollection(), ICWA.getCWAResModelElementCollections ()
Input Specification	The model elements of the model shown in Figure 1 are added at the IStorageAlter stub. Call getSatisfiesCWA() once. Call getCWAResInterfaces1() once. Call getCWAResInterfaces2() once. Call getSatisfiesCWAResRoleCollection() once. Call getCWAResModelElementCollections () once.

(Continued)

Output Specification	getSatisfiesCWA() returns False. getCWAResInterfaces1() returns a collection containing the GUIDs of i1 and i4. getCWAResInterfaces2() returns a collection containing the GUID of i1.
Output Specification	getSatisfiesCWAResRoleCollection() returns a collection containing the GUIDs of r1, r2 and r3 . getCWAResModelEleme ntCollections() returns an empty collection.
Environmental Needs	A stub that implements the ICWA interface. A stub that implements the IStorageAlter interface. A stub that implements IStorageRegistration.

(Other test cases are removed by the editor)

3.4 Event Router

Test Case Identifier	EventRouterT2
Test Item(s)	IEventRouter.evhDocumentCreated, IEventRouter. evhBeforeDocumentSave, IEventRouter. evhBeforeDocumentSaveAs & IEventRouter. evhBeforeDocumentClose.
Input Specification	EventRouter is correctly initialized. evhDocumentCreated(parDocumentA) is called once. evhBeforeDocumentSave(parDocumentA) is called once. evhBeforeDocumentSaveAs(parDocumentA) is called once. evhBeforeDocumentClose(parDocumentA) is called once.
Output Specification	An instance of the stub that replaces the TEventGateway class is created and properly initialized. eventGateway.evhBeforeDocumentSave(parDocumentA) is called twice. eventGateway.deref() is called once. eventGateway is destroyed.
Environmental Needs	An object parDocumentA implements the IVDocument interface and it is a valid iSpec document. An object parDocumentB that implements the IVDocument interface and is a UML document. A stub eventGateway that replaces the TEventGateway class.

(Other test cases are removed by the editor)

3.5 GUI

Test Case Identifier	GUIT1
Test Item(s)	IGUI.initialize, IGUI.deref
Input Specification	initialize(parStorageAlter, parStorageRetrieve, parVisio, parEventRegistration, parModelExplorer) is called once. deref() is called once.
Output Specification	No exceptions are raised.
Environmental Needs	A stub parStorageAlter that implements the IStorageAlter interface. A stub parStorageRetrieve that implements the IStorageRetrieve interface. A stub parVisio that implements the IVisio interface.

(Other test cases are removed by the editor)

3.6 TGUID

Test Case Identifier	TGUIDT1
Test Item(s)	TGUID.init, TGUID.toString, TGUID.compare
Input Specification	parString = 2287DC42-B167-11CE-88E9-0020AFDDD917. init(parString) toString parGUID.init(1136EB56 -A154-46DA-25C3-0020 AFDDD917); compare(me); compare(parGUID); parGUID.compare(me).
Output Specification	2287DC42-B167-11CE-88E9-0020AFDDD917; 0; an integer > 0; an integer < 0.
Environmental Needs	Another TGUID object parGUID.

3.7 TXMLParser

This component will be tested in a later stage, when other components can be used to simplify this testing.

4. Test Reports

4.1 Model Data Storage

Test report identifier	ModelDataStorageT1
Description	1) Initialize MDS; 2) Dereference MDS.
Activity and event retries	1) No exceptions are thrown.
Test report	Test successful.

4.2 Undo Assistant

The testing of this component will be done during the integration test, because the testing of this component is almost impossible without using other components.

4.3 Model Verifier

Test report identifier	ModelVerifierT1
Description	The CWA check is tested with a model with several violations to the CWA.
Activity and event retries	getSatisfiesCWA() returns False. getCWAResInterfaces1() returns a collection containing the GUIDs of i1 and i4. getCWAResInterfaces2() returns a collection containing the GUID of i1.
Test report	Test successful.

4.4 Event Router

Test report identifier	EventRouterT2
Description	1) Start session; 2) Create a new ISpec Document; 3) Save the ISpec Document; 4) Close the ISpec Document; 5) Open the ISpec Document; 6) Save the ISpec Document with a different name.

(Continued)

Test report identifier	EventRouterT2
Activity and event retries	1) Nothing happens; 2) A gateway initialized twice, dereferenced once; 3) The evhBeforeDocumentSave has been called once; 4) The evhBeforeDocumentSaveAs has been called once.
Test report	There is an error when throwing evhBeforeDocumentSave. The error has been corrected. After this the test is successful.

4.5 GUI

Test report identifier	GUIT1
Description	IGUI.initialize, IGUI.deref are tested.
Activity and event retries	No exceptions are thrown.
Test report	Test successful.

4.6 TGUID

Test report identifier	TGUIDT1
Description	TGUID.init, TGUID.toString, TGUID.compare are tested.
Activity and event retries	Output of to String was: 2287DC42-B167-11CE-88E9-0020AFDDD917. Output of compare was: 0, 1, -1.
Test report	This is within test specification, test successful.

4.7 TXMLParser

This component will be tested in a later stage, so there are no test results.

Post-reading exercises

Exercise 1

What are the main differences between low-level code modules and high-level code modules?

Exercise 2

Please connect the following modules with the level which they belong to.

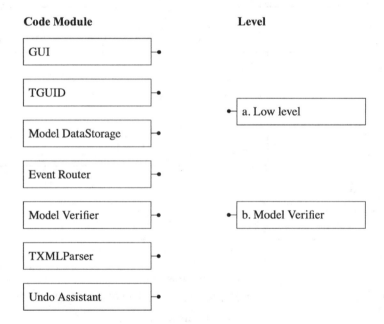

Exercise 3

Translate the following sentences into Chinese.

a) This document describes the Unit Test Plan (UTP) for the project and is written according to the software engineering standard provided by the General Computers Corporation.

_____

_____

b) The purpose of the unit tests is to make sure that the software units, or code modules, developed by the Interface Specification Tool project team comply with the requirements as stated in Chapter 4 of the ADD and Chapter 3 of the DDD.

c) The software to be tested consists of all the code modules developed for the IST Project.

d) A low-level code module normally is a single class, which is saved as a source file.

e) Before a high-level code module is to be unit tested, all corresponding low-level code modules have to be unit tested.

II E-mail & technical writing
Writing

🅐 Signing your e-mails

E-mails are now considered as an important means of communication. In the IT industry, where efficiency and brevity are appreciated, electronic communication has become increasingly popular because of its speed and broadcasting ability. Here is a series of guidelines to help you write effective professional e-mails.

Communicating Professionally and Effectively by E-mail

—Signatures

A "signature" is a small block of text appended to the end of your message, which usually contains your contact information. Many mailers can add their signature to your messages automatically. Signatures are a great idea but are subject to abuse. Balance is the key to a good signature.

√ **End your e-mails with your signature.**

Always close with your name, even though it appears at the top of the e-mail, and include contact information such as your phone, fax and street address. In many systems, particularly where mail passes through gateways, your signature may be the only means by which the recipient can identify who you are. More importantly, the recipient may want to communicate further by telephone, or send documents by post that cannot be e-mailed. Thus, including a formal signature block containing such useful information is preferred and professional.

Unit 8 Negotiating Assignments

√ **Include a complimentary closing before your signature.**

Ending your e-mail politely helps create the impression of you as a polite, professional businessperson. Try "Regards" or "Yours sincerely" or some other closing that you feel comfortable with and that is appropriate in the situation.

√ **Keep your signature short.**

Four to six lines is a handy guideline for maximum signature length. Signatures are perfect for conveying important information, but should remain to the point. Unnecessarily long signatures waste bandwidth (especially when distributed to a number of people) and can be annoying to the reader.

√ **Include your e-mail address and organization when writing to recipients outside the organization.**

If you do not put your e-mail address in the signature, sometimes it can be very difficult to locate your e-mail address, especially if it's forwarded to many people. If your e-mail address is a business address, you had better also include your title and company name in the signature. This is mandatory if you are representing your organization.

√ **Be cautious of fancy signatures.**

Some mailers allow you to attach additional random strings or images to your signature. This has become a fairly common practice. If you choose this option, it is recommended that the quote or image should reflect yourself. Keep it short and professional.

√ **Use the automated signature function.**

Most e-mail tools, such as Outlook, Thunderbird, Becky and Eudora, have an automated signature function. Using this function will attach your signature file to all outgoing messages. It allows you to create several alternative signature files as well. Make this function work for you.

× **Include excessive information when writing to colleagues.**

Normally, long signatures or signature files are not used in work-related e-mails, unless the e-mail is sent to recipients outside your organization. A closing

and a name may suffice as your signature internally.

× **Make the signature longer than seven lines.**

Lengthy signatures often annoy the recipients. In most situations, professionals frown on elaborate signatures and smile on appropriate ones.

× **Use embellished or silly signatures.**

Silly signatures with improper texts or images are definitely not recommended for business correspondence. Much of the humor and wit that is appropriate between friends or in a discussion group is out of place in the workplace.

B Exercises on e-mail writing

1. Study signatures from business correspondence on paper and in e-mails. Compare them. Are they different? How? Compose an e-mail to your instructor about your findings.

2. Imagine you are writing an e-mail to your partner, who wants to know more about unit testing. In your e-mail, you need to:

- briefly introduce unit testing;
- describe a common procedure of unit testing;
- ask him/her to reply if there are any further questions.

Send this e-mail to your partner and your instructor, and review the message written by your partner.

C Guide to technical writing

No matter what your current or future job is, writing will be essential to your work because you will have to communicate your technical knowledge to others. Technical documents are a most frequently used type of writing in the IT industry, so training in technical writing will help ensure a smooth career path. Here is a series of guidelines to help you write successful technical documents.

Guidelines to Successful Technical Writing

—Choosing the right words and expressions

Your choice of words ultimately determines the quality of your writing. Keep your expression simple, jargon-free, original, convincing, precise, concrete and specific.

Replace difficult words and phrases with simpler alternatives.

Flowery diction and needless jargon obscure your message and force your readers to work too hard for understanding. Word it in plain English, and avoid inflated diction. Do not use three syllables when one will do. Here is a list of a number of words and expressions that should generally be avoided in favor of the simple alternative.

approximately	=	about
ascertain	=	find
assist, assistance	=	help
commence	=	start
demonstrate	=	show
dwelling	=	house
effectuate	=	do
(to) endeavor	=	(to) try
endeavor	=	effort
enquire	=	ask
facilitate	=	help
in consequence	=	so
in excess of	=	more
in respect of	=	about
in the event of	=	if
initiate	=	begin
multiplicity of	=	many
necessitate	=	need
phenomenon	=	event

terminate	=	end, stop
transmit	=	send
utilize	=	use
Owing to the situation that...	=	Because, since...
Should a situation arise where...	=	If...
Taking into consideration such factors as...	=	Considering...

Also, unless you are discussing building maintenance or computer graphics, never use the verb "render".

Faulty   The testing strategy rendered it impossible to find all the faults.

Correct   The testing strategy made it impossible to find all the faults.

In other words, if you mean "make", then write "make" not "render".

Count the syllables and trim when you can. The following is an example of bad diction inflation:

Inflated   Re-evaluate the design of the entire user interface to minimize design factors which are resulting in sensitive and/or critical maintenance and inspection procedures.

Revised   Redesign the UI so they are easier to maintain and inspect.

Use verbs instead of nouns if possible.

Look at the following sentence:

Half the team was involved in the development of system Y.

This sentence contains a classic example of a common cause of poor writing style. The sentence is using the abstract noun "development" in place of the verb "develop". The simpler and more natural version of the sentence is:

Half the team was involved in developing system Y.

Turning verbs into abstract nouns always produces sentences longer than necessary, so avoid doing it. The following examples show the improvement you can achieve by replacing nouns with verbs:

Faulty   He used to help in the specification of new software.

Unit 8 Negotiating Assignments

Revised He used to help specify new software.

Faulty Clicking the icon causes the execution of the program.

Revised The program executes when the icon is clicked.

Faulty The analysis of the software was performed by Fred.

Revised Fred analyzed the software.

Faulty It was reported by Jones that method Z facilitated the utilization of inspection techniques by the testing team.

Revised Jones reported that method Z helped the testing team use inspection techniques.

Use jargon only if it helps you communicate better.

Expressions like MS/DOS, UI, ODBC and glitch are examples of jargon. In general, jargon refers to a special vocabulary of a particular group or activity. It is often shorthand or abbreviations, with specialized usage. If you are confident that every reader of your report understands the specialty, then it may be used. For example, if your only potential readers are computer specialists, it is appropriate to say that a computer is down without having to explain what "down" means. In all other cases (which are almost always) jargon should be avoided. If you cannot avoid using such expressions, then define the term the first time you use it or refer the reader to a glossary where it is defined.

Needless jargon Intercom utilization will be used to initiate overtime programmer operative involvement.

Revised Programmers who have to work overtime will be notified on the intercom.

Be consistent in naming the same subject or object.

The rule "Never use the same word twice." does not apply to all forms of writing. Some people may feel they must use different words to describe the same thing. In technical writing the opposite rule applies: "You should always use the same word to refer to the same thing." Failing to do so may confuse and annoy readers.

Consider, for example, the following paragraph that was written in a group project final report:

In the first three weeks of the project we wrote a project plan for the system. We were ambitious in our requirements because we wanted the group project to be a success and we wanted the software to be of high quality. In fact, we were determined that our software would be very satisfactory. By the end we realized there were major problems with the project. The first increment of the project we delivered was inconsistent with the requirements specification and it was clear the final code would not be the best system as there were clearly better groups than ours.

The problem with this paragraph is that there are three key objects that are referred to in different and inconsistent ways. The objects are:

The project:	It refers to the entirety of the group experience.
The plan:	It refers to a document describing the requirements and the schedule for implementing them.
The system:	It refers to the software system that the group project is supposed to deliver.

Unfortunately, we find that these things are referred to in different parts of the paragraph as:

The project:	project; group project; group.
The plan:	project plan; requirements; requirements specification.
The system:	system; software; project; code; final code.

In situations such as this, it is important to identify each different object first and decide once and for all how it should be named. Once you have made this decision consistently, use the same name throughout when you refer to that object. Applying this instruction to the example above will yield the following improved text:

In the first three weeks of the project we wrote a plan for the system. Our plan was ambitious because we wanted the project to be a success and we wanted the system to be high quality. In fact, we were determined that our project would be very satisfactory. By the end we realized there were major problems with the project. The first increment of the system we delivered was inconsistent with the plan and it was clear the final system would not be the best system as there were clearly better projects than ours.

D Technical writing exercises

Exercise 1

Improve the economy and directness of the following sentences by replacing difficult words and phrases with simpler alternatives, and replacing nouns with verbs.

a) Do not hesitate to contact us in the event that you are in need of assistance for the computer software application at any time.

b) Bill made the suggestion that we hire an additional systems analyst.

c) We request the formation of a committee of experienced software engineers for the review of quality discrepancies.

Exercise 2

Revise the following paragraph to make it more internally consistent and readable.

Good software engineering is based on a number of key principles. One such principle is for the group to get a good understanding of the customer requirements. It is also important to deliver in regular increments, involving the client as much as possible. Another rule is that it is necessary to do unit testing, black box and white box testing throughout, with unit testing being especially crucial. In addition to the previous doctrines, a programmer needs to be able to maintain good communication within the software team (and also with the user).

Exercise 3

Imagine you are designing a test case for the GUI of a program developed by you. Fill in the prepared form with your design.

[SYSTEM] [Module] [Item]

Test Scenario Specification Form

Version No.:	Build #:	Requirement #:	Page:
Release No.:	Test Scenario #:	Environment:	of
		Server Tested:	

Written By	Date	Reviewed By	Date	Executed By	Date	Retry #

Instructions:

Test Scenario Objective:

Assumptions/Dependencies:

Test Files/Test Data:

STEP#	DESCRIPTION	EXPECTED RESULT	ACTUAL RESULT

Unit 9

Testing Software

软件测试

I Software test plans

II E-mail & technical writing

Software test plans
Reading

A An overview

Software Test Plan

A software test plan is a document describing the scope, approaches, resources, and schedule of intended testing activities. It identifies test items, the features to be tested, the testing tasks, who will do each task, and any risks requiring contingency planning.

Why do we need test plans?

Even the simplest software system is so complex and is so prone to failure, that planning to test is just as essential as planning your design.

The second major reason is that the IT industry is not doing a good job at releasing quality products. Currently tested and released software averages more than ten significant bugs per 1000 lines of code. Thinking ahead and planning your testing is one way to cut that down.

Who should we execute test plans?

It is essential to have a different person or team execute the Test Plan other than those who developed the software. Individuals who like to break things are most suitable for this kind of job. It is also helpful to develop a creative tension between development teams who test each other's software.

Unit 9 Testing Software

B A sample of software test plan

Read the following software test plan. This is only a preliminary document, but it sheds much light on what a software project test plan is. For the first time, please only scan the whole document. Keep these questions in mind and try to answer them after scanning. Time limit: 10 minutes.

- What's the primary purpose of the tests mentioned in this test plan?
- How many functions are to be tested?
- What dependencies will the test team need in order to fully conduct the tests?

General Computers Corporation Corporate Payroll System Project

Test Plan
Revision C
Revision History

DATE	REV	AUTHOR	DESCRIPTION
5/14/xx	A		First Draft
5/21/xx	B		Second Draft
5/25/xx	C		Added FTBT

Glossary

corporate *adj.* 团体的, 公司的

payroll *n.* 薪酬

Table of Contents

1. Introduction
    1.1 Test Plan Objectives
2. Scope
    2.1 Data Entry
    2.2 Reports
    2.3 File Transfer
    2.4 Security

3. Test Strategy

   3.1 System Test

   3.2 Performance Test

   3.3 Security Test

   3.4 Automated Test

   3.5 Stress and Volume Test

   3.6 Recovery Test

   3.7 Documentation Test

   3.8 Beta Test

   3.9 User Acceptance Test

4. Environment Requirements

   4.1 Data Entry Workstations

   4.2 Mainframe

5. Test Schedule

6. Control Procedures

   6.1 Reviews

   6.2 Bug Review Meetings

   6.3 Change Request

   6.4 Bug Reporting

   6.5 Bug Report Form

7. Functions to Be Tested

8. Resources and Responsibilities

   8.1 Resources

   8.2 Responsibilities

9. Deliverables

10. Suspension/Exit Criteria

11. Resumption Criteria

12. Dependencies

Glossary

recovery *n.* 恢复, 复原

mainframe *n.* 主机, 特大型机

suspension *n.* 暂停

resumption *n.* 继续, 重新开始

Unit 9　Testing Software

　　12.1 Personnel Dependencies
　　12.2 Software Dependencies
　　12.3 Hardware Dependencies
　　12.4 Test Data & Database
13. Risks
　　13.1 Schedule
　　13.2 Technical
　　13.3 Management
　　13.4 Personnel
　　13.5 Requirements
14. Tools
15. Documentation
16. Approvals

> **Glossary**
> approval *n.* 核定, 批准
> outgrow *v.* 过大而不适于

1. Introduction

The corporation has outgrown its current payroll system & is developing a new system that will allow for further growth and provide additional features. The software test department has been tasked with testing the new system.

The new system will do the following:

Provide the users with menus, directions & error messages to direct him/her on the various options;

Handle the update/addition of employee information;

Print various reports;

Create a payroll file and transfer the file to the mainframe;

Run on the corporate Intranet using IBM compatible PCs as data entry terminals.

1.1 Test Plan Objectives

This Test Plan for the new payroll system supports the following objectives:

Define the activities required to prepare for and conduct System, Beta and User Acceptance Tests;

Communicate the System Test strategy to all responsible parties;

Define deliverables and responsible parties;

Communicate the various Dependencies and Risks to all responsible parties.

2. Scope

2.1 Data Entry

The new payroll system should allow the payroll clerks to enter employee information from IBM compatible PC workstations running Windows 98 or higher. The system will be menu driven and will provide error messages to help direct the clerks through various options.

2.2 Reports

The system will allow the payroll clerks to print three types of reports. These reports are:

A pay period transaction report;

A pay period exception report;

A three-month history report.

Glossary

clerk *n.* 职员

2.3 File Transfer

Once the employee information is entered into the LAN database, the payroll system will allow the clerk to create a payroll file. This file can then be transferred to the mainframe over the network.

2.4 Security

Each payroll clerk will need a user ID and password to log in to the system. The system will require the clerks to change the password every 30 days.

3. Test Strategy

The test strategy consists of a series of different tests that will fully exercise the payroll system. The primary purpose of these tests is to uncover the system's limitations and measure its full capabilities. A list of the various planned tests and a brief explanation follows below.

3.1 System Test

The system tests will focus on the behavior of the payroll system. User scenarios will be executed against the system as well as screen mapping and error message testing. Overall, the system tests will test the integrated system and verify that it meets the requirements defined in the requirements document.

3.2 Performance Test

Performance test will be conducted to ensure that the payroll system's response times meet the user's expectations and do not exceed the specified performance criteria. During these tests, response

> **Glossary**
>
> limitation *n.* 限制, 缺陷

times will be measured under heavy stress and/or volume.

3.3 Security Test

Security tests will determine how secure the new payroll system is. The tests will verify that unauthorized user access to confidential data is prevented.

3.4 Automated Test

A suite of automated tests will be developed to test the basic functionality of the payroll system and perform regression testing on areas of the systems that previously had critical/major bugs. The tool will also assist us by executing user scenarios with several emulated users.

3.5 Stress and Volume Test

We will subject the payroll system to mass input conditions and a high volume of data during the peak times. The system will be stress tested using twice (200 users) the number of expected users.

3.6 Recovery Test

Recovery tests will force the system to fail in various ways and verify the recovery is properly performed. It is vitally important that all payroll data is recovered after a system failure and no data corruption occurs.

3.7 Documentation Test

Tests will be conducted to check the accuracy of the user documentation. These tests will ensure that no features are missing, and the contents can be

Glossary

unauthorized *adj.* 未经认可的

confidential *adj.* 机密的

regression *n.* 蜕变（测试）

emulated *adj.* 模拟的

vitally *adv.* 极其地

corruption *n.* 损坏

occur *v.* 发生，出现

Unit 9 Testing Software

easily understood.

3.8 Beta Test

The Payroll Department will beta test the new payroll system and report any bugs they find. This will subject the system to tests that could not be performed in our test environment.

3.9 User Acceptance Test

Once the payroll system is ready for implementation, the Payroll Department will perform User Acceptance Testing. The purpose of these tests is to confirm that the system is developed according to the specified user requirements and is ready for operational use.

4. Environment Requirements

4.1 Data Entry Workstations

20 IBM compatible PCs (10 will be used by the automation tool to emulate payroll clerks);

Pentium III processor (minimum);

128-Megabyte RAM;

100-Megabyte Hard Drive;

Windows 98 or higher;

Attached to the corporate Intranet;

A network attached printer;

20 user IDs and passwords (10 will be used by the automation tool to emulate payroll clerks).

4.2 Mainframe

Attached to the corporate Intranet

Access to a test database (to store payroll information transferred from LAN payroll system)

5. Test Schedule

- Ramp-up/System familiarization
                                    6/01/xx–6/15/xx
  System Test   6/16/xx–7/26/xx
  Beta Test   7/28/xx–8/18/xx
  User Acceptance Test   8/29/xx–9/03/xx

6. Control Procedures

6.1 Reviews

The project team will perform reviews for each phase (i.e. Requirements Review, Design Review, Code Review, Test Plan Review, Test Case Review and Final Test Summary Review). A meeting notice, with related documents, will be e-mailed to each participant.

6.2 Bug Review Meetings

Regular weekly meeting will be held to discuss reported bugs. The development department will provide status/updates on all bugs reported and the test department will provide additional bug information if needed. All members of the project team will participate.

6.3 Change Request

Once testing begins, changes to the payroll system are discouraged. If functional changes are required, these proposed changes will be discussed with the Change Control Board (CCB). The CCB will determine the impact of the change and if/when it should be implemented.

> **Glossary**
> ramp-up *n.* 稳步增加（测试）
> familiarization *n.* 熟悉
> discourage *v.* 不鼓励
> impact *n.* 影响

Unit 9 Testing Software

6.4 Bug Reporting

When bugs are found, the testers will complete a bug report on the bug tracking system. The bug tracking system is accessible by testers, developers and all members of the project team. When a bug has been fixed or more information is needed, the developer will change the status of the bug to indicate the current state. Once a bug is verified as FIXED by the testers, the testers will close the bug report.

Glossary

indicate v. 指明, 说明

6.5 Bug Report Form

VHTNSoftware Development   Problem Report   #: _____

Program _____   Release _____   Version _____

Report Type (1-6)_____   Severity(1 -3)_____   Attachments   (Y/N)_____
1 - Coding error   1 - Fatal   If yes, describe:
2 - Design issue   2 - Serious   _____
3 - Suggestion   3 - Minor   _____
4 - Documentation
5 - Hardware
6 - Query

Problem Summary   _____

Can you reproduce the problem? (Y/N)   _____

Problem & How can it be reproduced?
_____
_____
_____

Suggested fix (optional input)
_____
_____
_____

        Reported By:_____   Date: _____

           Items below are for use only by the development team

Functional Area: _____   Assigned To:_____

Comments:
_____
_____
_____

Status _____   Prioirity(1-5)_____
1 - open   2 - closed

Resolution(1-9)_____   Resolution Version _____
1 - Pending   4 - Deferred   7 - Withdrawn by reporter
2 - Fixed   5 - As designed   8 - Need more info
3 - Irreproducible   6 - Can't be fixed   9 - Disagree with suggestion

Resolved By: _____   Date: _____

Tested By: _____   Date: _____

Treat as Deferred   (Y/N)_____

7. Functions to Be Tested

The following is a list of functions that will be tested:

Add/update employee information;

Search/look up employee information;

Escape to return to Main Menu;

Security features;

Scaling to 700 employee records;

Error messages;

Report printing;

Creation of payroll files;

Transfer of payroll files to the mainframe;

Screen mappings (GUI flow), including default settings;

FICA calculation;

State tax calculation;

Federal tax calculation;

Gross pay calculation;

Net pay calculation;

Sick leave balance calculation;

Annual leave balance calculation.

A Requirements Validation Matrix will "map" the test cases back to the requirements. See Deliverables.

> Glossary
>
> matrix *n.* 矩阵
> coordinate *v.* 协调，调整

8. Resources and Responsibilities

The Test Lead and the Project Manager will determine when system test will start and end. The Test Lead will also be responsible for coordinating schedules, equipment, and tools for the testers as well as writing/updating the Test Plan, Weekly Test Status Reports and Final Test Summary Report. The testers will be responsible for writing the test cases and executing the tests. With the help of the Test Lead, the Payroll Department Manager and payroll clerks will be responsible for the Beta and User Acceptance Tests.

8.1 Resources

The test team will consist of:

A Project Manager;

A Test Lead;

5 testers;

The Payroll Department Manager;

5 payroll clerks.

8.2 Responsibilities

Project Manager	Responsible for project schedules and the overall success of the project. Participate in the CCB.
Lead Developer	Serve as a primary contact/liaison between the development department and the project team. Participate in the CCB.
Test Lead	Ensure the overall success of the test cycles. Coordinate weekly meetings and communicate the testing status to the project team. Participate in the CCB.
Testers	Responsible for performing the actual system testing.
Payroll Department Manager	Serve as liaison between the Payroll Department and the project team. Help coordinate the Beta and User Acceptance testing efforts. Participate in the CCB.
Payroll Clerks	Assist in performing the Beta and User Acceptance testing.

Glossary

liaison *n.* 联络

9. Deliverables

Deliverable	Responsible Parties	Completion Date
Develop test cases	Testers	6/11/xx
Test case review	Test Lead, Dev. Lead, Testers	6/12/xx
Develop automated test suites	Testers	7/01/xx
Requirements validation matrix	Test Lead	6/16/xx

Unit 9 Testing Software

(Continued)

Deliverable	Responsible Parties	Completion Date
Obtain user IDs and passwords for payroll system/database	Test Lead	5/27/xx
Execute manual and automated tests	Testers, Test Lead	8/26/xx
Complete bug reports	Everyone testing the product	On-going
Document and communicate test status/coverage	Test Lead	Weekly
Execute Beta Tests	Payroll Department Clerks	8/18/xx
Document and communicate Beta Test status/coverage	Payroll Department Manager	8/18/xx
Execute User Acceptance Tests	Payroll Department Clerks	9/03/xx
Document and communicate Acceptance Test status/coverage	Payroll Department Manager	9/03/xx
Final Test Summary Report	Test Lead	9/05/xx

10. Suspension/Exit Criteria

If any bugs are found which seriously impact the test progress, the QA manager may choose to suspend testing. Criteria that will justify test suspension are:

Hardware/software is not available at the times indicated in the project schedule.

Source code contains one or more critical bugs, which seriously prevent or limit testing progress.

Assigned test resources are not available when needed by the test team.

11. Resumption Criteria

If testing is suspended, resumption will only occur when the problem(s) that caused the suspension has been resolved. When a critical bug is the cause of the suspension, the "fix" must be verified by the test department before testing is resumed.

12. Dependencies

12.1 Personnel Dependencies

The test team requires experienced testers to develop, perform and validate tests. It will also need the following resources available: application developers and payroll clerks.

12.2 Software Dependencies

The source code must be unit tested and provided within the scheduled time outlined in the project schedule.

12.3 Hardware Dependencies

The MainFrame, 10 PCs (with specified hardware/software) as well as the LAN environment need to be available during normal working hours. Any downtime will affect the test schedule.

12.4 Test Data & Database

Test data (mock employee information) & database should also be made available to the testers for use during testing.

13. Risks

13.1 Schedule

The schedule for each phase is very aggressive and could affect testing. A slip in the schedule in one of the other phases could result in a subsequent slip in the testing phase. Rigorous project management is crucial to meeting the forecasted completion date.

13.2 Technical

Since this is a new payroll system, in the event of a failure the old system should be able to run. We will run our test in parallel with the production system so that there is no downtime of the current system.

13.3 Management

Management support is required in order that when the project falls behind, the test schedule does not get squeezed to make up for the delay. Management can reduce the risk of delays by supporting the test team throughout the testing phase and assigning people to this project with the required skills set.

13.4 Personnel

Due to the aggressive schedule, it is very important to have experienced testers on this project. Unexpected turnovers can impact the schedule.

> **Glossary**
>
> aggressive *adj.* 积极上进的
> slip *n.* 失误, 误差
> subsequent *adj.* 随后的
> crucial *adj.* 至关重要的
> forecast *v.* 预测, 事先安排
> turnover *n.* 人员更替
> attrition *n.* 人员损耗

If attrition does happen, all efforts must be made to replace the experienced individual.

13.5 Requirements

The Test Plan and test schedule are based on the current Requirements Document. Any changes to the requirements could affect the test schedule and will need to be approved by the CCB.

14. Tools

The Acme Automated test tool will be used to help test the new payroll system. We have the licensed product on site and installed. All of the testers have been trained on the use of this test tool.

15. Documentation

The following documentation will be available at the end of the test phase:

Test plan;

Test cases;

Test case review;

Requirements validation matrix;

Bug reports;

Final test summary report.

16. Approvals

Name (Print)   Signature   Date

1) _____

2) _____

3) _____

C Post-reading exercises

Exercise 1

What kinds of risks may the test team encounter throughout the testing phase?

Exercise 2

Here is a list of the various tests in the plan. Try to use as few words as possible to give a brief explanation for each test.

System Test: _____.
Performance Test: _____.
Security Test: _____.
Automated Test: _____.
Stress and Volume Test: _____.
Recovery Test: _____.
Documentation Test: _____.
Beta Test: _____.
User Acceptance Test: _____.

Exercise 3

Translate the following sentences into Chinese.

a) The corporation has outgrown its current payroll system and is developing a new system that will allow for further growth and provide additional features.

b) Once the employee information is entered into the LAN database, the payroll system will allow the clerk to create a payroll file.

c) The primary purpose of these tests is to uncover the system's limitations and measure its full capabilities.

d) Overall, the system tests will test the integrated system and verify that it meets the requirements defined in the requirements document.

e) Performance test will be conducted to ensure that the payroll system's response times meet the user's expectations and do not exceed the specified performance criteria.

11 E-mail & technical writing
Writing

A Dealing with attachments appropriately

E-mails are now considered an important means of communication. In the IT industry, where efficiency and brevity are appreciated, electronic communication has become increasingly popular because of its speed and broadcasting ability. Here is a series of guidelines to help you write effective professional e-mails.

Communicating Professionally and Effectively by E-mail

—Attachments

√ **Attach a file to your e-mail only when you have to.**

Most e-mail applications allow you to attach almost any type of file to an e-mail message, including word processor documents, spreadsheets, sound files, and graphics. Send such files only when they are necessary, otherwise they may inhibit the recipient from accessing and acting on your main message in a timely fashion. They may also impair the performance of the mail servers. In cases where you must include them, compress multiple files into one, especially photos, when possible.

√ **Specify your attachments in your message.**

Use proper labeling conventions and file extensions to identify your documents. Within your e-mail message, specify that this message has an attachment. Identify the attachment by its proper file name, the application software version, and the content description and size of the attached file, so the recipient can quickly determine how to access it. Keep in mind that your recipients may not have the same suite of desktop application software you have, or if they do have the same application, they may not have the same version of software as your attachment. If possible, you should always check with the recipient to verify the compatibility of the attachment being sent. If you are unable to verify compatibility, send the file in an earlier version (Word 97 vs. Word 2003) or a more common format, i.e. text, RTF or HTML, to increase the likelihood that the file can be read by the greatest number of recipients.

√   **Protect your recipients from viruses.**

Sending and reading e-mail messages do not pose a threat of catching a computer virus. Viruses can only be contained in attachments to your e-mail messages. Be very careful about your attachment files. Run an anti-virus program that checks the attachment before you send it.

×   **Attach unnecessary files to your e-mail.**

Put your information in the body of your e-mail whenever possible. In most instances, you can copy and paste the relevant text into the e-mail rather than send the entire word processor file (unless of course your recipient actually needs to view file in order to edit or archive it). Consider including Uniform Resource Locators (URLs) or pointers to FTP (File Transfer Protocol) versions. Sending unnecessary attachments to many people is a significant waste of digital resources.

×   **Forget the attachment when you hit the "Send" button.**

Many people receive e-mails with a missing attachment. If you consistently make this mistake, people (perhaps important clients) may be unimpressed. They may even hesitate to do business with you in the future. When you get ready to send your e-mail, ask yourself: "What am I forgetting?"

×   **Keep attachments as small as possible.**

Attaching large documents, files, images or programs may make your message so large that it cannot be delivered or will consume excessive resources. A good rule of thumb is to not send a file larger than one megabyte. Note that e-mail providers may restrict the size of message a user is able to receive. When you send a large attachment, the receiver may be locked out of his/her mailbox.

B Exercises about e-mails and software defects

1. Consult with IT professionals on the job or in your own company. Inquire if they send files mainly by e-mail, and what they think about sending bulk e-mail attachments. Are they aware of the rules governing e-mail attachments as discussed above? Write your findings in an e-mail memo, and send the memo to your instructor.

2. Imagine you have discovered a new bug in a piece of software your group is testing. You want to write to the Test Manager about it. In your e-mail:

- Tell the Manager that a *Save As* dialogue came up when you double-clicked the status bar;
- Suggest a possible reason;
- Ask him/her if it's OK to attach the bug report along with your e-mail;
- Ask him/her to contact you if there are any further questions.

Send this e-mail to your partner and your instructor, and review the message sent from your partner.

C Guide to technical writing

No matter what your current or future job is, writing will be essential to your work because you will have to communicate your technical knowledge to others. Technical documents are a most frequently used type of writing in the IT industry, so training in technical writing will help ensure a smooth career path. Here is a series of guidelines to help you write successful technical documents.

Guidelines to Successful Technical Writing
—Grammar, punctuation & mechanics

Units 7 and 8 explain the most important principles for improving the style of your writing. However, it is also important (and actually easier) to improve the grammar, punctuation and mechanics of your writing. No matter how vital and informative a message may be, its credibility is damaged if it contains errors. Any errors—illogical, fragmented, or run-on sentences; faulty punctuation; poorly chosen words—stand out and mar otherwise good writing.

This section provides several guidelines to avoid the most commonly made language errors. Although it offers no easy solution to long-standing fundamental problems in writing, it does provide a simple guide for basic improvement. You may discover that some writing problems are easier to solve than you had realized.

Grammar

Be aware of dangling modifiers.

A dangling modifier, a common case of English misusage, is a verbal phrase, prepositional phrase, or dependent clause that does not refer to the subject in its sentence. Correct it by rewriting the sentence to make the modifier correspond with the subject.

Dangling   After a night of hard work, the bug was finally fixed.

Revised   After we had worked all night, the bug was finally fixed.

Dangling   Without knowing the final design, it is difficult to make plans for the implementation.

Revised   Because we have not received the final design, it is difficult to make plans for the implementation.

Dangling   The project ended up a failure, not having studied the user requirements carefully.

Revised   We could not complete the project, not having studied the user requirements carefully.

Avoid run-on sentences.

Run-on sentences cram too many ideas into one sentence without providing needed breaks or pauses between thoughts.

Run-on The hourglass is more accurate than the water clock for the water in a water clock must always be of the same temperature in order to flow with the same speed since water evaporates it must be replenished at regular intervals thus not being as effective in measuring time as the hourglass.

Revised The hourglass is more accurate than the water clock because water in a water clock must always be of the same temperature to flow at the same speed. Also, water evaporates and must be replenished at regular intervals. These temperature and volume problems make the water clock less effective than the hourglass in measuring time.

Avoid faulty subject and verb agreement.

Failure to make the subject of a sentence agree in person and number with the verb is a common writing error. Luckily, it is an error easily avoided or corrected. In short sentences, where the subject and the verb are not far apart, this mistake is not likely to occur; however, in more complicated sentences, you may lose track of the subject-verb relationship.

Faulty Everyone in the development group and the testing group *have* worked long hours.

Correct Everyone in the development group and the testing group *has* worked long hours.

Faulty The high number of software projects that failed this year are disappointing.

Correct The high number of software projects that failed this year is disappointing.

Faulty Computer security techniques and a workable plan for evading various kinds of hacker attacks *requires* large expenditures.

Correct   Computer security techniques and a workable plan for evading various kinds of hacker attacks *require* large expenditures.

Avoid sentence shifts.

Shifts in point of view damage coherence. If you begin a sentence or paragraph with one subject or person, remain with it.

Shift in person	When you have managed to write professional English technical documents, one will have a great sense of achievement.
Revised	When you have managed to write professional English technical documents, you will have a great sense of achievement.
Shift in voice	He delivered the specs for system design, and the UML diagrams were also revised by him.
Revised	He delivered the specs for system design, and also revised the UML diagrams.
Shift in number	One should send the project manager a progress report before they are off work.
Revised	One should send the project manager a progress report before one is off work.
	Or
	Send the project manager a progress report before getting off work.

Punctuation

Use semicolons with adverbs as conjunctions.

You must use semicolons, not commas, to accompany adverbs and other expressions that connect related independent ideas. Here are some common adverbs: *besides, otherwise, still, however, furthermore, moreover, consequently, therefore, on the other hand, in contrast, in fact*. For example:

The project was finally completed; however, the customers were not very satisfied.

Designing such a complicated system is too demanding; in fact, no one is sure if we can make it within the time limit.

Use colons only after an independent clause.

The colon introduces explanations or lists.

She is an ideal colleague: honest, reliable, and competent.

Do not place a colon directly after a verb.

Faulty　　Our plans include analyzing the requirements, hiring two new programmers, getting enough funds, and getting the work done.

Use hyphens to avoid ambiguity.

re-creation *(a new creation)*

recreation *(leisure activity)*

Mechanics

Consider your audience before using abbreviations.

Do not use abbreviations that might confuse your reader. Often, abbreviations are not appropriate in formal writing. When in doubt, write out the full words.

Correct　　Check component No. 3.

Faulty　　Mr. Zhang, our specialist consultant, is a Dr.

Always avoid abbreviating words out of laziness.

For example: never write "approx." for "approximately" (it may be better to write "about"); never write "e.g." for "for example".

D Technical writing exercises

Exercise 1

Locate an English technical document. Study the document and ask yourself if it conforms to the principles discussed in the guide above. If not, how could you improve its grammar, punctuation, and mechanics? Discuss your conclusion in class or write it in one or two paragraphs.

Exercise 2

The following sentences need to be revised. Make necessary changes and be prepared to give reasons for each.

a) Reforming a design specification is beyond our means, we have no time left for that.

b) Jane wonders if Mr. Lee has received her document and like it?

c) A programmer should be very skilled at coding, on the other hand, he/she also has to be trained in testing.

d) The product was finally delivered in the A.M. yesterday.

e) Although nearly finished, we left the meeting early because we were worried about our progress.

Exercise 3

Organize into groups of four, and gather bug information from online bugzillas. Analyze it and fill in the following form with a typical bug.

Unit 9   Testing Software

VHTNSoftware Development   Problem Report   #: _____

Program _____ Release _____ Version _____

Report Type (1-6) _____   Severity(1 -3) _____   Attachments   (Y/N) _____
1 - Coding error   1 - Fatal   If yes, describe:
2 - Design issue   2 - Serious   _____
3 - Suggestion   3 - Minor   _____
4 - Documentation
5 - Hardware
6 - Query

Problem Summary _____

Can you reproduce the problem? (Y/N)   _____

Problem & How can it be reproduced?
_____
_____
_____

Suggested fix (optional input)
_____
_____
_____

        Reported By: _____   Date: _____

Items below are for use only by the development team

Functional Area: _____   Assigned To: _____

Comments:
_____
_____

Status _____   Prioirity(1-5) _____
1 - open   2 - closed

Resolution(1-9) _____   Resolution Version _____
1 - Pending   4 - Deferred   7 - Withdrawn by reporter
2 - Fixed   5 - As designed   8 - Need more info
3 - Irreproducible   6 - Can't be fixed   9 - Disagree with suggestion

        Resolved By: _____   Date: _____

        Tested By: _____   Date: _____

Treat as Deferred   (Y/N) _____

Unit 10 Closing Off

项目总结

I   Post-mortem reports
II Summaries

Post-mortem reports
Reading

A An overview

Post-mortem Analysis

After a software project is completed it enters the post-mortem phase. During this phase the project manager and the project team should get together and carry out a post-mortem review.

The post-mortem review looks back at the project with the primary aim of providing feedback to future projects. The following is an indicative list of some of the important information that should come up in the post-mortem analysis:

- Comparison of original planned dates with actual completion dates;
- Significant reasons for deviations between the planned and actual performances;
- Original planned cost and effort versus actual cost and effort analysis;
- Analysis of recording of work to determine the ratios of times spent in various activities;
- Analysis of the code review findings to determine the trend of defects with respect to time for each team member;
- Analysis of the code review findings to indicate which type of defects are predominant, so that appropriate training can be given to the team members;
- Come up with metrics such as defects found per thousand lines of code, number of limits coded per day, total number of lines of code, and functions.

Very few projects go as planned. Many projects encounter problems that must be corrected and a few lucky projects go smoother than planned. Regardless

Unit 10 Closing Off

of how successful or disastrous a project is, it is important to review the project in detail once the project is over. This allows your team to figure out what things were done well and to document the things that need improvement. It also aids in building a knowledge base that teams coming behind you can review to ensure they get the most out of their upcoming projects.

B A sample of post-mortem report

Read the following post-mortem report. This is only a preliminary document, but it sheds much light on what a post-mortem report is. For the first time, please only scan the whole document. Keep these questions in mind and try to answer them after scanning. Time limit: 10 minutes.

- What's the primary purpose of this post-mortem report?
- Where shall we refer to know the main objectives of this project?
- How many main lessons can be learned from this project?

National Science and Technology Promotion Program in ICT Engineering

The Development and Implementation of an Integrated Web and Voice Based Public Services System for Rural Community in China, an E-Government Pilot Project

Post-mortem Report

Executive Summary

This GCC project is intended to develop and implement an Integrated Web- and Voice-based Public Services System for rural community in China. The project accomplishment has fulfilled the initial project plan, both in the project deliverables and time/budget consideration. The actual cost of the project is USD

Glossary

ICT = Information & Communications Technology
rural *adj.* 农村的
community *n.* 区域, 社区
pilot *adj.* 示范的, 试验的
accomplishment *n.* 成就
fulfill *v.* 实现, 完成

37,338.16 or about 69.5% of the total budget approved by the central government. The <u>collaboration</u> program between developers from GCC and the central government has also met the project objectives, especially in enhancing their ICT <u>competencies</u> and adding their project experiences. Several input and lessons learned have also been considered for further development and collaboration. In addition, the main result of this project, voice-based public system, is currently in operation in several Chinese rural communities. This service has provided <u>significant</u> benefits for those public communities, especially in providing easier access to the on-line government information services, as well as in enhancing their ICT <u>awareness</u>. Finally, this system will be <u>promoted</u> as a model of voice-based public services for most local governments in China.

Glossary
collaboration *n.* 协作
competency *n.* 能力
significant *adj.* 重要的, 重大的
awareness *n.* 意识
promote *v.* 推广

Document Status Sheet

Document Title	Post-mortem Report
Author(s)	David Zhang, Chris King, Jacky Chen
Version	0.4
Document Status	internally accepted

Version	Date	Author(s)	Summary
0.1	07-29-20xx	David Zhang	Document created before post-mortem meetings.

Unit 10 Closing Off

(Continued)

0.2	08-16-20xx	David Zhang, Jacky Chen	Revised document after collecting post-mortem analyses from the development team.
0.3	08-25-20xx	David Zhang, Chris King, Jacky Chen	Added comments from the test team.
0.4	08-26-20xx	David Zhang, Chris King, Jacky Chen	Revised the Lessons Learned section.

Glossary

obstacle *n.* 障碍, 阻碍

encounter *v.* 遭遇, 遇到

Document Change Record

Document Title	Post-mortem Report
Date of Changes	08-26-20xx
Section Number	Reason for change
4	More lessons added.

1. Introduction

1.1 Purpose

The purpose of this Post-mortem Report is to review the team's work in this project and to describe in detail the specific activities that are most effective and those that need adjustments prior to the next project. One goal of the document is to evaluate the accomplishments done by the whole team. Another goal is to inform future project teams of the obstacles encountered during this release.

This document will focus upon identifying the following:

Project summary;

Accomplishments;

Lessons and action items (actions that can be taken to improve future projects).

1.2 Scope

The second chapter provides a summary of the project, including an introduction, objectives, functions, budgets, activities involved, and collaborations. The third chapter lists the accomplishments and positive impacts. The fourth chapter presents the learned lessons and actions for further developments.

1.3 List of Definitions

ADD	Architectural Design Document
ICT	Information and Communications Technology
USD	United States Dollars
eGIS	e-Government Information System
V-eGIS	Voice-based e-Government Information System
IVR	Interactive Voice Response

1.4 List of References

ADS	Architectural Design Specification
DDS	Detailed Design Specification
UTR	Unit Test Report
ITR	Integration Test Report
STR	System Test Report
PTR	Performance Test Report
SeTR	Security Test Report
FTR	Field Trial Report

2. Project Summary

2.1 Overview

This project has been selected by the Ministry of Information Industry and GCC as a National Science and Technology Promotion Program in ICT Engineering. This project commenced on January 21, 20xx and should be completed by the end of August, 20xx. The total budget approved for the project is USD 53,753.50.

This project was conducted by General Computers Corporation and Ace Telecom. Other affiliations supporting this project are the Communications Research Laboratory, the University of Science, and the University of Electronic Communications.

The two main activities for this project are the development of the Integrated Web- and Voice-based Public Services System, and the field trial of the system which has been successfully implemented in several rural communities in China, since August 1, 20xx.

2.2 Objectives

The project objectives are:

a) Developing a web-based public service system "e-Government Information System (eGIS)", which can be accessed by public service machines installed in downtown areas or by web browsers.

b) Implementing the system in more developed rural areas in South China.

c) Developing a voice-based public service system which can be accessed by using an ordinary telephone. This application is named Voice-eGIS or V-eGIS. The V-eGIS developed in this project will be linked to the existing eGIS.

d) Implementing the V-eGIS in less developed rural areas for enhancing the public services from the local governments.

To give better access for other rural areas, which don't have Internet access, eGIS will be enhanced using telephone-based access (V-eGIS), so people in less developed rural areas may get similar information using ordinary phones. V-eGIS application is developed based on the common business processes of local governments in China by using the Web and telephone as interfaces, with an easy navigation menu which helps user with activities related to the existing process flows.

The project can be considered as a model in bridging the digital gap in rural China by giving public community access to e-Government public services using web and voice technologies.

2.3 Functions of the System

The developed web-based public service system, eGIS, is a comprehensive management system for public services provided by local governments in several more developed rural areas in South China. eGIS has three main functions: payment in a cashier, services and monitoring.

V-eGIS will cover three main functions: payment in a cashier, services and monitoring, with an additional function for the tracking status of the applicant. In this system, an applicant will request his/her permit from Public Service Office. After that, he/she can check the permit status using telephone. IVR (interactive voice response) server that is connected to the Public Service real-time

Glossary

comprehensive *adj.* 综合的
cashier *n.* 出纳处, 交款处
permit *n.* 许可, 凭证

database performs the status checking process using voice. The officers from the government also can monitor the applicant's status the same way the applicant does.

2.4 Financial Statement

Summary of the financial statement is described in the following table.

- Fund transferred by the government

    = USD 35,453.00

- Actual cost of the project = USD 37,338.16

- Balance = USD (1,885.16)

Item	Budget (USD)	Actual (USD)	Variance (USD)
Travel Expense	18,865.00	11,927.58	6,937.42
Daily Allowance	8,886.00	6,208.00	2,678.00
Others	18,948.00	17,200.00	1,748.00
Correspondence	6,054.50	1,976.00	4,078.50
Shipping	1,000.00	26.58	973.42
Total	53,753.50	37,338.16	16,415.34

> **Glossary**
> balance n. 结余, 余额
> allowance n. 津贴, 补助

A detailed account of this financial statement is described in the Final Accounting Report.

2.5 Project Activities

In general, all main activities of the project complied with the initial planning. Although some activities required a slight time adjustment, the project closure was exactly similar to the initial plan, which is August 30, 20xx.

The detailed description about project activities is summarized in the following table.

No	Activity	Plan		Actual		Summary
		Schedule	Place	Schedule	Place	
1	System Requirements Analysis	01/19/xx-02/23/xx	City A	01/30/xx-02/04/xx	Town B	Due to administrative difficulties, the development of the System Specifications was not really being held in City A. This discussion was held in Town B and the developers from City A Government involved in this project participated in this discussion via Internet and audio (telephone). The output of this activity: System Requirements Specification of the eGIS and V-eGIS.
2	Procurement	01/26/xx-02/27/xx	Town B	01/30/xx-02/18/xx	Town B	The output of this activity: preparing all equipment required for this project.
3	Network Design	01/26/xx-01/28/xx	Town B	01/26/xx-02/28/xx	Town B	This activity was supposed to design the network configuration of eGIS and V-eGIS, including the design of the latter's connectivity to the former database.
4	Software Development					
	·Design	01/26/xx-02/30/xx	Town B	01/26/xx-03/04/xx	Town B, City A	The initial activity for developing the software design drafts was primarily in Town B. The initial drafts were finalized in City A.
	·Coding	03/01/xx-03/27/xx	Town B	03/07/xx-04/16-xx	Town B	This activity intended to develop the software based on the design documents developed previously.

Unit 10 Closing Off

(Continued)

No	Activity	Plan		Actual		Summary
		Schedule	Place	Schedule	Place	
5	Site Survey	-	-	03/03/xx-03/05/xx	Town C	Site Survey was required to make sure all equipment and software application developed in this project are fit to the existing systems. The Project Team was also presenting the concept of the eGIS and V-eGIS Systems to executives of the rural communities.
6	Software & Hardware Installation	02/28/xx-03/14/xx	Town B	04/18/xx-04/23/xx	Town C	Regarding the initial plan, this activity would be in Town B. For efficiency, this actual activity had been moved directly to the trial site (Town C).
7	Software Integration	03/03/xx-03/04/xx	Town B	04/21/xx-04/23/xx	Town C	This activity integrated all eGIS and V-eGIS modules.
8	Testing I	03/30/xx-06/02/xx	Town B	04/22/xx-06/24/xx	Town B	Activity of Testing I covered functional test, performance test and security test of eGIS and V-eGIS systems, especially regarding the flow of the software. Dr. Wang's team from University of Science accomplished this test remotely from Town B. In addition, Dr. Wang suggested some technical considerations, including the security system of eGIS and V-eGIS.

(Continued)

No	Activity	Plan		Actual		Summary
		Schedule	Place	Schedule	Place	
9	Testing II	03/30/xx-07/22/xx	Town C	04/24/xx-07/29/xx	Town C	This is another test done by two test teams from GCC. The tests covered integration test and system test. The test results concluded that the two software systems were error free and ready for field trial.
10	Field Trial	07/23/xx-08/10/xx	Town C	07/30/xx-08/10/xx	Town C	GCC test team performed test and stabilization tests. This field trial concluded that the system was stable and ready for launching.
11	Training Admins	-	-	08/15/xx-08/20/xx		This activity was intended to train the network administrators from the rural governments to be ready for operation and maintenance of eGIS and V-eGIS after launching.
12	Launching	08/29/xx	Town C	08/29/xx	Town C	The launching of eFIS and V-eGIS was performed by Town C Mayor, and Mr. Zhang from GCC. This activity was also viewed as the formal closure of this grand project.

2.6 Collaborations

Generally, all project activities were actively supported by all engineers involved in this project. Some activities were virtually held via a project mail list (egis@yahoo.com) and via telephone line and other activities required a physical attendance of the engineers.

In this case, the activities that required the attendance of engineers from the partners are:

1) Software Design

In this activity, four Ace Telecom engineers were in City A from February 28 to March 4, for finalizing the system specifications and software designs. Prof. Jason Gibson, from the University of Electronic Communications, hosted this session. In addition, Prof. Gibson also arranged an e-application benchmark program.

2) Testing-I

Dr. Wang, expert from the University of Science, visited Town B from May to June to clarify the system specifications, network configurations, and functional test of eGIS and V-eGIS. Dr. Wang provided suggestions regarding the specification and design, especially in security matters. Dr. Wang also provided a lecture for GCC engineers about *"The Advanced Technology on Wireless and Smart-card Security"* on May 23.

3) Installation & Field Trial

Two engineers from the Communications Research Laboratory visited Town C in April and July, for installation and field trials of eGIS and V-eGIS.

4) Systems Launching

Two Ace Telecom engineers attended this launching. This activity was also viewed as the formal project closure.

Glossary

benchmark *n.* 基准测试

3. Accomplishments

3.1 Project Outputs

The main outputs of this project are:

a. Software Application eGIS

The software application of eGIS is completed and ready to be implemented in cities and towns in China that had no e-Government service system. eGIS will be sufficient for some more developed rural areas, while for other less developed areas, V-eGIS will be implemented in further steps.

b. Software Application V-eGIS

The software application of V-eGIS is completed and ready to be implemented in less developed cities and towns in China that have already had existing e-Government services.

c. Establishing eGIS and V-eGIS Services for Town C

The project has been successful in the introduction and implementation of the web- and voice-based e-public services in Town C. People can access this service via installed service machines, web browsers, or telephones. With this Mandarin service, people of Town C can easily access information and monitor the status of their services provided by government.

All equipment (server, IVR card, networking element, etc.) required for the public service pilot project in Town C is provided by this project.

d. Competency Improvement of Human Resources

Glossary

sufficient *adj.* 足够的, 充分的

All engineers involved in this project obtain some advantages to improve their technical capabilities and large project experiences. In addition, some related human resources of local governments are also trained in operation and maintenance of this web- and voice-based e-public services.

3.2 Project Sustainability

China has targeted that as much as 50% of the local governments should implement e-Government services in the next few years. It implies that the market of e-Government in China is very promising in the long term. The demand for delivering clean and transparent services to the public is fully essential for achieving better quality and improvement in human resources, business environment and local facilities. On the other hand, telephone penetration in China is relatively bigger than that of the Internet, so the sustainability of the project will be in the long run.

3.3 Impact Assessment

Some of the more important impacts in relation to key points in the project's objectives are:

a. Established a transparent, effective and efficient system of public services, especially for people who reside in small cities and rural areas;

b. Provided an easier access to the on-line information services by using telephones;

c. Widened the coverage area of e-Government implementation;

d. Increased IT awareness from community,

Glossary

sustainability *n.* 可持续性
promising *adj.* 有前途的
transparent *adj.* 透明的
penetration *n.* 普及度

especially in Town B and Town C;

e. The same model and design of the eGIS systems can be extended to other public service systems such as e-police and e-health for public community.

4. Lessons Learned

The lessons learned from this project are as follow:

a. The original schedule failed in correctly estimating the time for design and testing. Therefore the project team had to spend less time in testing than expected. Luckily, the unit tests were quite efficient, enabling the integration and system tests to be more compact, where the test results were also acceptable. Time estimation in schedule can be more flexible and realistic.

b. Due to failures of communications facilities, some engineers didn't manage to attend certain significant meetings. Collaboration among engineers from some different locations can be effectively accomplished electronically via both Internet and telecommunication facilities.

c. In developing a voice-based application system for local government in China, it would be more effective and more influential for the public to use multi-language voice menus or instructions. The model of multi-language menus provides an option for people with international backgrounds. This

> **Glossary**
>
> compact *adj.* 紧凑的，简短的
>
> influential *adj.* 有影响力的

Unit 10 Closing Off

feature should be considered for the next release.

  d. The software applications for public services should be designed with stronger security system because they contains sensitive and secret data. For this end, next-generation eGIS and V-eGIS should ensure that the data be encrypted in the database server using a strong application such as CipherGate. Furthermore, an additional function to <u>randomize</u> the registration number and authentication procedure will also be considered in the future releases.

Glossary

randomize v. 使随机, 随机化

C Post-reading exercises

Exercise 1

  How many main functions does V-eGIS have? What are they?

Exercise 2

  Please connect the following activities of the project with their related descriptions.

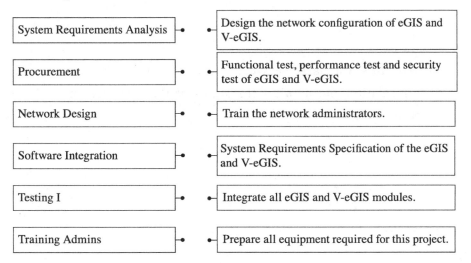

System Requirements Analysis	↔	Design the network configuration of eGIS and V-eGIS.
Procurement	↔	Functional test, performance test and security test of eGIS and V-eGIS.
Network Design	↔	Train the network administrators.
Software Integration	↔	System Requirements Specification of the eGIS and V-eGIS.
Testing I	↔	Integrate all eGIS and V-eGIS modules.
Training Admins	↔	Prepare all equipment required for this project.

Exercise 3

Translate the following sentences into Chinese.

a) This GCC project is intended to develop and implement an Integrated Voice- and Web-based Public Services System for rural community in China.

b) The project accomplishment has fulfilled with the initial project plan, both in the project deliverables and time/budget consideration.

c) In addition, the main result of this project, voice-based public system, is currently in operation in several Chinese rural communities.

d) The project can be considered as a model in bridging the digital gap in rural China by giving public community access to e-Government public services using web and voice technologies.

e) With this Mandarin service, people of Town C can easily access information and monitor the status of their services provided by government.

II Summaries
Writing

A Summary of guidelines to professional e-mails

E-mails are now considered an important means of communication. In the IT industry, where efficiency and brevity are appreciated, electronic communication has become increasingly popular because of its speed and broadcasting ability. Here is a series of guidelines to help you write effective professional e-mails.

Communicating Professionally and Effectively by E-mail

—Summary

Electronic mail is fast becoming a primary means of communication but because it is still a relatively new communications medium, conventions as to how it should be used are not well developed. This passage summarizes the do's and don'ts of e-mail writing covered in this text book. The purpose is to help us to be more efficient in using this medium. This summary can also serve as a final checklist for you to review your daily e-mail usage.

Address an e-mail correctly.

√   Always provide a Personal Name if your mail system allows it.

√   Put only the people you are directly addressing in the "To" field.

√   Put the people you are indirectly addressing in the "Cc" field.

√   Use "Bcc" when addressing a message that will go to a large group of people who do not necessarily know each other.

×   Use impolite or inexplicit personal names.

- ✗ Use "Cc" to copy your message to everyone.

Write meaningful subject lines.

- √ Always include a subject line in your message.
- √ Try to use a subject that is meaningful to the recipient as well as to yourself.
- √ If you are replying to a message but are changing the subject of the conversation, change your entry in the subject line as well.

Open polite and clear salutations in e-mails.

- √ Use appropriate salutations.
- √ Use titles only when you are sure what they are.
- √ Use familiar and simple salutations if it is appropriate to do so.
- √ Begin the body of e-mails with greetings only when you have to.
- √ Identify yourself.
- √ Put your identity at the beginning of an e-mail.
- ✗ Use simple salutations for people with formal traditions.

Use adequate length, content and format.

- √ Keep the message readable.
- √ Stay on the subject as much as possible.
- √ Get your points in order.
- √ Use correct grammar and spelling.
- √ Use appropriate and professional language in your communications.
- √ Format messages for easy reading.
- √ Use plain text most of the time.
- ✗ Type your message in all-uppercase.
- ✗ Include confidential information in your e-mails.
- ✗ Send public "flames"—messages sent in anger.
- ✗ Use HTML or RTF formats as often as possible.

Unit 10 Closing Off

×	Ask to recall a message.

Pay attention to your manners and tones.

√	Mind your manners and use socially courteous expressions frequently.
√	Use a positive, friendly and appropriate tone.
√	Maintain professionalism.
√	Tailor your message for the receiver.
×	Write or send e-mails that will backfire.
×	Write what you would not say.

Reply and forward e-mails properly.

√	Every message, except spam or junk mail, deserves a reply.
√	Give your final answer in your reply.
√	Respond as soon as possible.
√	When replying, include enough of the original message to provide a context.
√	When replying or forwarding, minimize the original message.
√	Distinguish between text quoted from the original message and your reply.
√	Pay careful attention to where your reply or forward is going to end up.
×	Reply to e-mails with one-word answers or questions.
×	Forward e-mails unless you have the permission of the author.
×	Forward forwarded messages on to your friends and co-workers.

Use appropriate abbreviations and smileys.

√	Learn the most common abbreviations.
√	Make sure the recipients understand the abbreviations.
√	Learn the most common smileys.
√	Use smileys to avoid misunderstandings.
×	Overuse abbreviations that cause frustrations.
×	Overuse emoticons that damage your image.

Sign your e-mails professionally.

√ Remember to end an e-mail with your signature.

√ Always add a complimentary closings before your signature.

√ Keep your signature short.

√ Include your e-mail address and organization when writing to external recipients.

√ Avoid using fancy signatures.

√ Make good use of the automated signature function.

× Include excessive information when writing to colleagues.

× Make the signature longer than seven lines.

× Use embellished or silly signatures.

Deal with e-mail attachments properly.

√ Attach a file to your e-mail only when you have to.

√ Specify your attachments in your message.

√ Protect your recipients from viruses.

√ Avoid unnecessary attachments.

× Forget the attachment when you hit the "Send" button.

× Send large attachments.

B E-mail about course summary

Write to your instructor about what you have learned about e-mail in this course, and feel free to give any suggestions or complaints about the text, additional materials, exercises or instructors. Write keeping in mind the rules you have learned.

C Guide to technical writing

No matter what your current or future job is, writing will be essential to your work because you will have to communicate your technical knowledge to others. Technical documents are a most frequently used type of writing in the IT industry, so training in technical writing will help ensure a smooth career path. Here is a series of guidelines to help you write successful technical documents.

Guidelines to Successful Technical Writing

—Summary

After nine units, we have covered a lot of topics in technical writing. It is time for a summary.

In technical writing, you report specialized information for the practical use of readers who have requested it. Recipients of your e-mail may need your information to perform a task, answer a question, solve a problem, or make a decision.

Here is a summary of guidelines for successful technical writing. It can also serve as a final checklist for your technical documents:

Prewriting

Decide what the objective of the document is.

Write down the objective.

Have a specific reader in mind.

Decide what information you need to include.

Have access to a good dictionary.

Identify someone who can provide feedback.

Outlining

Informal Outlines

List all the relevant topics in any order at first.

Identify the major categories of related information from your list.

Arrange the information categories in a meaningful order.

Formal Outlines

Make the list of major topics broad enough to encompass your subject.

Also make the outline specific enough so you can discuss each topic in detail. Use logical notation and consistent formatting.

Make all items of equal importance parallel, or equal, in grammatical form. Present the topics in a logical sequence.

Be sure that all items in your outline meet the objectives of the document.

Drafting

Topic Sentences

Write a topic sentence for a paragraph.

Before you can write a good topic sentence, you must identify your purpose, based on what you know of your readers' needs.

The idea you make in your topic sentence must be narrow enough to be covered in one paragraph.

Ways to Develop a Paragraph

Provide examples of the topic.

Include facts, statistics, evidence, details, or precedents that confirm the topic. Quote, paraphrase, or summarize the evidence of other people on the topic.

Describe an event that has some influence on the topic.

Define terms, if necessary, that are related to the topic.

Explain how something works.

Describe the physical appearance of an object, area, or person.

Technical Descriptions

Review your prewriting before you draft.

Reorganize your outline.

Preface your text with a title precisely stating the topic of your description. Write a topic statement.

Write an introduction.

Draft the text to the best of your ability.

Write your conclusion according to your purpose.

Short Reports

Write in good organization.

Develop your report properly.

Adopt a correct language style.

Letter Reports

Follow the standard letter format.

Do not forget to insert additional headings where appropriate.

Be sure your letter report retains the personal "you" perspective.

Memo Reports

Write a heading with the organization name, sender, recipient, subject, and date.

If the memo report is longer than one page, then list the recipient, date and page number on following pages.

Most memos simply end after the final point.

Progress Reports

Make sure your progress report answers the most essential questions.

Include the topic about which you are reporting and the reporting interval in the subject.

Bring your readers up-to-date with background data or a reference to previous reports.

Choose an optimum organization and format.

Prepared-form Reports

Write entries as requested on prepared forms .

Make your language simple and clear for quick review.

Attach your own statement explaining certain items on the form if needed.

Planning Proposals

Use appropriate format and supplements.

Write a concrete and specific title.

Start with a clear introduction.

Develop a plain and understandable body.

Design your planning proposal to reflect your attention to detail.

Focus on one specific subject and remain with it.

Identify all problems readers themselves might not recognize.

Provide concrete and specific information.

Use visuals whenever possible.

End with a comprehensive summary.

Revising

Sentences

Revise unclear sentences and avoid ambiguity.

Use active rather than passive voice most of the time.

Avoid unnecessary words and repetition.

Make sentences fluent.

Words and Expressions

Replace difficult words and phrases with simpler alternatives.

Use verbs instead of nouns if possible.

Use jargon only if it helps you communicate better.

Be consistent in naming the same subject or object.

Unit 10 Closing Off

Grammar, Punctuation and Mechanics

Be aware of dangling modifiers.

Avoid run-on sentences.

Avoid faulty subject and verb agreement.

Avoid sentence style shifts.

Use semicolons with adverbs as conjunctions.

Use colons only after an independent clause.

Use hyphens to avoid ambiguity.

Take into account your audience before you use abbreviations.

Always avoid abbreviating words because of laziness.

D Technical writing exercises

Exercise 1

Gather all the writings you produced in previous units. Check them against the guidelines above. Do they need improvement? Which of them are better than the others? Which of them are worse? Revise and edit the one that in your view needs the most revision.

Exercise 2

Write a final report about your IT English training. Write this report both as a summary of your work during the training and as a demonstration of writing final reports that comply with high standards of communication as to their organization, format, content, and style.